The FORD
8 H.P. CHASSIS

Instructions for Dismantling & Reassembling.

Issued by Facilities Department

FORD MOTOR COMPANY LIMITED

DAGENHAM · ESSEX

FORD Y & C MODEL REGISTER

This publication has been assembled by using scans of original pages from 'The Ford 8 H.P. Chassis' issued by the Facilities Department of Ford Motor Company, Dagenham Essex, in 1932, *and with their permission to reproduce.*

NOTE: This book was produced for the early models produced, (SR) and many changes were made in later production. (LR, C, CX)

It should be noted that many blank pages (originally put in for making notes) have been omitted.

Page numbers shown on illustrations on the bottom edge are those in the original.

Illustrations and photos have often been expanded to fit the paper size. This book, written in the early days, may not reflect many improvements and changes made since printing.

Reference should be made to "Henry's Car For Europe" by Sam Roberts as well as the service bulletins available from:

Ford Y & C Model Register.
www.fordyandcmodelregister.co.uk

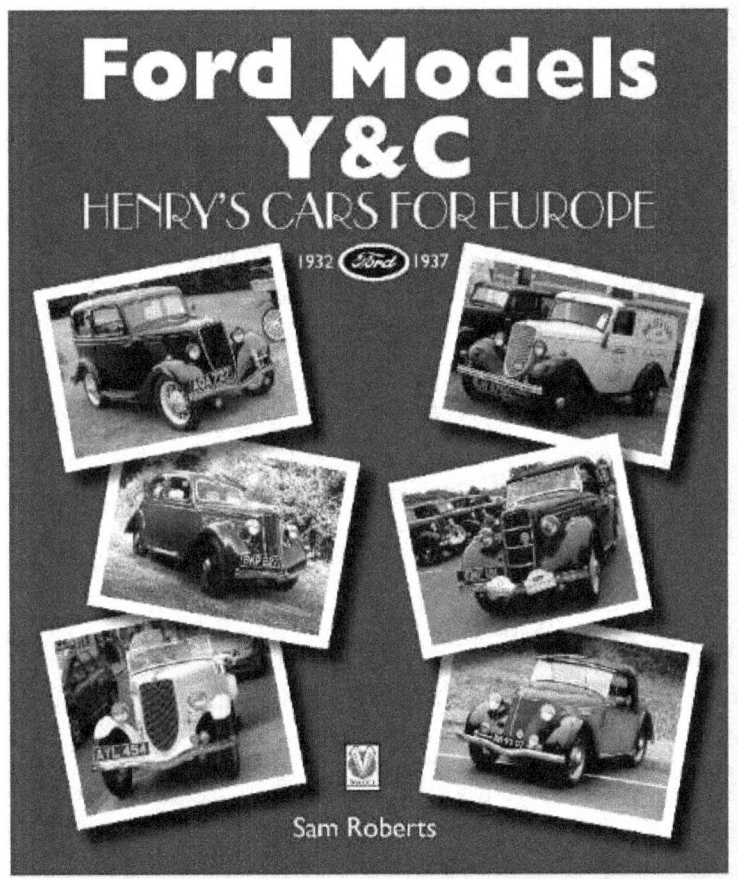

**THE DEFINITIVE HISTORIES OF THE
8hp FORD MODEL Y & 10hp MODEL C of 1932 to 1937**
and all their worldwide variants, complete with detailed technical specifications and production records. Most of the great names of the Ford Motor Company were involved with these small cars. Contains 270 pages and 300 photographs in a hardback cover.

Introduction

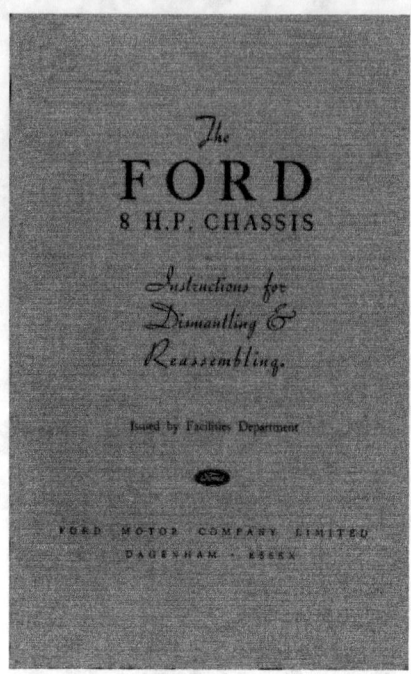

This book has been carefully transcribed and reprinted in this form, taken from an old copy of what is known as "The Blue Book" – a rare and hard-to-find copy that was issued by the Ford Motor Company to schools and colleges to aid instruction to new mechanics coming into the flourishing auto industry. Ford also donated demonstration chassis to colleges on which to practice these techniques described.

Most auto engineers and mechanics had previously been self-taught in cycle makers or blacksmithies, and Ford was keen to encourage the trade to support sales and maintenance of their new models.

The general layout and presentation style has been kept, but to aid reading and understanding, certain tabs and indentations may have been brought in line with modern practice.

Some illustrations may have been changed to modern photos, to improve quality with better resolution, and extra illustrations also added along with explanatory notes.

Some terms used e.g. "Wrench" (Spanner) have been changed to more normal terms. Many such changes that are extra to the original are shown in *different* text or font style to differentiate them.

Many special tools mentioned are rare and difficult to find, so alternatives or other methods may be noted (in other text or font style.)

The aim of this exercise is to allow modern-day enthusiasts to better understand how to maintain their cars, be it a Model Y, C, CX or any of the numerous variants still in use today.

This should be used in conjunction with the original handbooks from those days, the 'Bulletins' issued to garages, and proprietary handbooks by Pitmans and others. Reprints of some are available from the Ford Y & C Model Register to members

Copyright ©2022 Roger Corti

This edition Published February 2023

To join our Register, scan this QR code and follow link

Manual of Instruction

FOR

DISMANTLING & REASSEMBLING

THE

8 h.p. Ford Chassis

Issued by Facilities Department

Ford Motor Company Ltd.

DAGENHAM, ESSEX

Foreword

THE high speed of modern transport and the place it occupies in modern civilization demands that the younger generation should be given an opportunity of acquiring knowledge of this important subject as a part of their education.

This manual has been arranged to provide a practical course on the 8 h.p. Ford which would appeal to educational authorities for use in Technical and Secondary Schools. It describes the operations involved in dismantling and rebuilding the chassis, and details the special equipment which permits these operations to be carried out with the maximum efficiency and minimum delay.

A text book covering elementary mechanical principles and their development in the modern high efficiency motor car has been specially compiled to serve as a companion volume to this manual. By means of these two volumes, a sound theoretical and practical understanding of modern motor car design will readily be acquired in a simple and progressive manner.

Contents

Preliminary Rules 7

Section 1
ENGINE AND CLUTCH 9-61

Section 2
GEAR BOX 63-91

Section 3
BACK AXLE 93-125

Section 4
FRONT AXLE 127-149

Section 5
STEERING GEAR 151-173

Section 6
DISTRIBUTOR, GENERATOR, STARTER MOTOR, CARBURETTOR, FUEL PUMP 177-205

Each of the above Sections is divided into four parts, namely:

- A. Removing unit from chassis.
- B. Dismantling unit.
- C. Reassembling unit.
- D. Installing unit in chassis.

Preliminary Rules

1. **Lay out special tools and equipment** and arrange them conveniently to hand, the smaller tools on floor tray 142.

2. **Cover front mudguards** to prevent chipping and scratching of enamel surface.

3. **Disconnect Battery Negative Lead** from electrical system to avoid accidental short circuits and starting of engine. (For correct procedure see Section 1A, Operation 1).

4. **Place all major parts of the car** as they are removed on parts carrier 398, and all smaller parts with nuts, bolts, etc., in partitioned tray 344. (See illustrations at beginning of Sections 1 and 3).

Notes

TOOLS—At the beginning of each division of this manual, the special tools and equipment required are listed with their respective numbers. A practical knowledge of their use and application will readily be gained as work progresses under the guidance of the manual.

PARTS—All chassis parts are described and many are illustrated in the Parts Catalogue, which should be used as a cross reference to this manual. With the help of this catalogue the student will quickly become familiar with all parts and associate with each its correct name.

Part numbers may be found to vary in prefix and suffix between this manual and the parts catalogue. This will indicate a slight change in design of that part.

Section I

ENGINE AND CLUTCH

A. To remove Engine and Clutch from chassis.

B. To dismantle Engine and Clutch.

C. To re-assemble Engine and Clutch.

D. To install Engine and Clutch in chassis.

ILLUSTRATIONS AND MEMORANDA

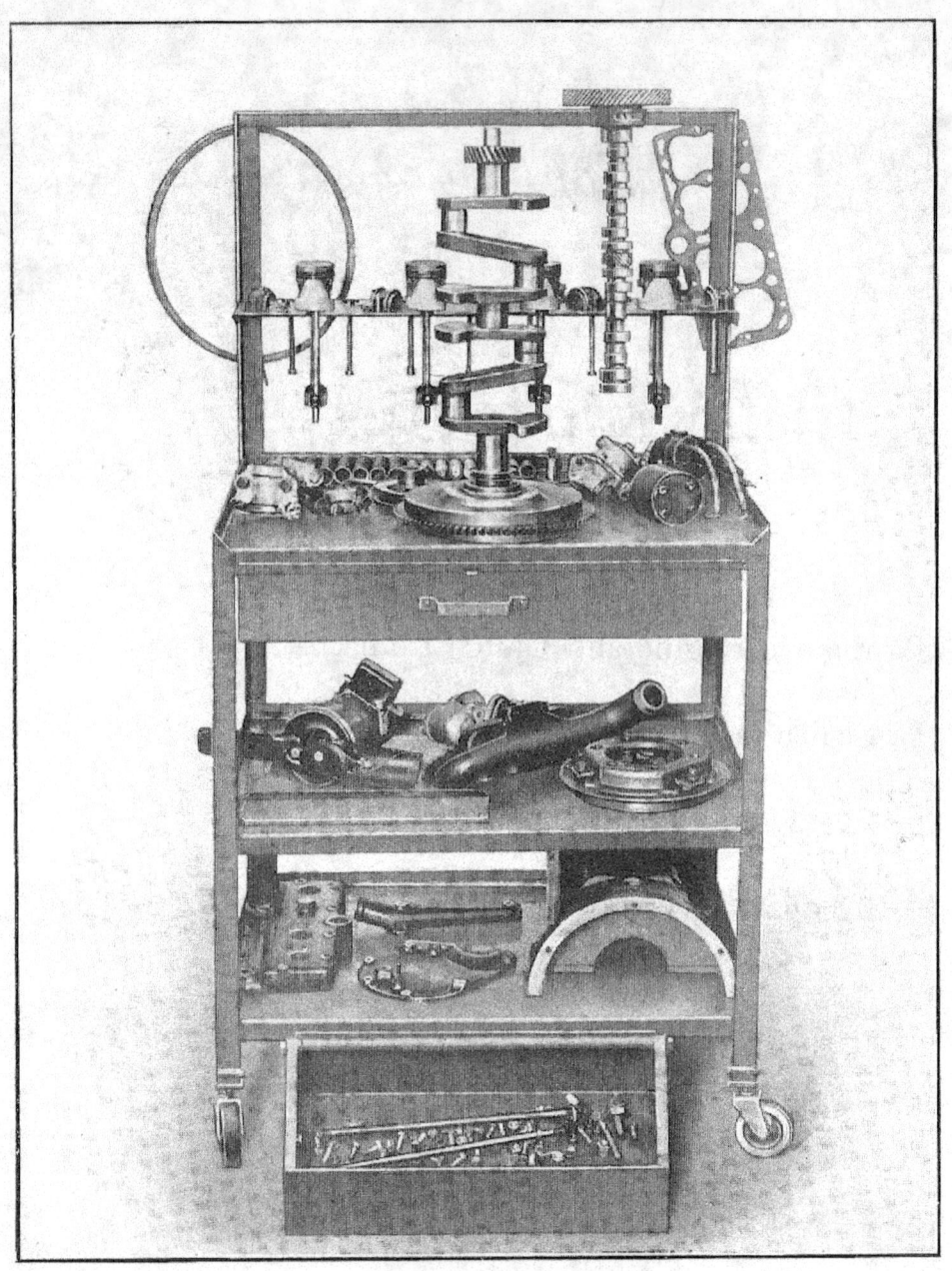

TO REMOVE ENGINE AND CLUTCH FROM CHASSIS

Special Tools and Equipment Required

Tools from Standard Tool Kit

Double ended wrench ($\tfrac{7}{16}''$ and $\tfrac{1}{2}''$)	B-17015
,, ,, ,, ($\tfrac{9}{16}''$ and $\tfrac{5}{8}''$)	B-17016
Sparking plug and cylinder head nut wrench	Y-E-17017
Screw-driver	B-17020

Special Tools and Equipment

Engine stand	AB-35
,, ,, adapter	Y-36
,, lifting eye bolt	Y-46
Lifting hoist	73
Creeper	76
Copper hammer	83
Two drain pans	127
Floor tray	142
Partitioned tray	344
Parts carrier	398
Wrench for windshield wiper connection gland nut, coil terminal and starting motor	Y-810
Sparking plug spanner	Y-859
Wrench	1616
Hexagon socket	2120
Wrench	2242
Adapter	2291
Cap for gear box cover	—
Three dummy sparking plugs	48-M-416
Special plug for flexible pipe	—
1 pair mudguard covers	—

Fig. 1.

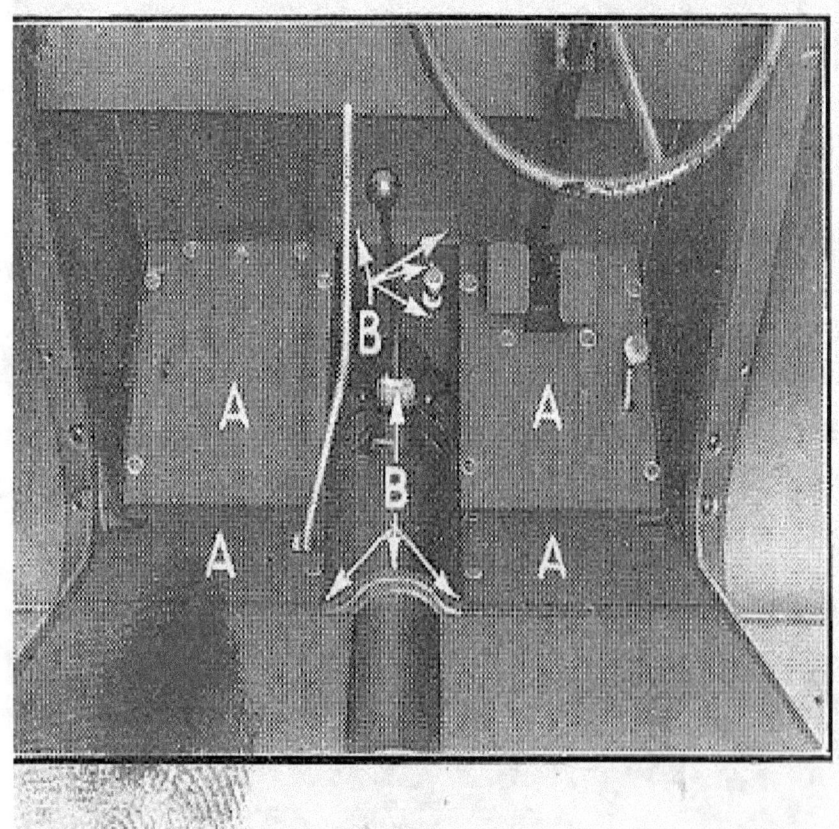

Fig. 2.

TO REMOVE ENGINE AND CLUTCH FROM CHASSIS

Carry out in sequence the following operations:—

Operation		Illustration
1.	Place protective covers over front mudguards. Lift off-side of bonnet and disconnect battery by removing two cover fastening nuts Y-110932 with screw-driver B-17020, slacking off battery negative terminal clamp nut with wrench B-17015 and lifting terminal off battery post. Replace battery cover and nuts loosely.	1A
2.	Place drain pan 127 under engine sump and remove drain plug B-6730 with wrench YE-17017.	
3.	Place drain pan 127 under radiator and turn on drain cock B-8115 which is located on radiator bottom tank at the off-side and may be found by reaching up, under the front splash shield.	
4.	Remove screw and nut holding off-side of rear bonnet hinge bracket Y-16729 to scuttle with screw-driver B-17020. Close off-side of bonnet.	
5.	Open near-side of bonnet. Free radiator brace rod Y-8132-B at front end by slacking off rear nut with wrench B-17015. Remove screw and nut holding near-side of rear bonnet hinge bracket Y-16729 to scuttle with screw-driver B-17020 and lift off bonnet. Remove radiator brace rod from bracket on radiator.	
6.	Remove eight screws holding off side No. 1 (sloping) floor board YF-940130-B in position with screw-driver B-17020. Slacken lock nut holding accelerator pedal pad YE-11471 in position with wrench B-17015, screw off pedal pad and lift out No. 1 floor board.	2A

Operation	Illustration

6(*cont.*) Remove three screws holding off-side No. 2 (horizontal) floor board YF-940132-B in position and lift out floor board. Remove six screws holding near-side No. 1 floor board YF-940131-B in position and lift out floor board. Remove three screws holding near-side No. 2 floor board YF-940133-B in position and lift out floor board.

7. Remove two screws holding foot dimmer switch Y-110378 to floor board plate YF-940016 with screw-driver B-17020 and push switch down.
Remove two screws holding front of floor board plate to dash.
Remove three screws holding rear of floor board plate to cross member.
Slack off locking ring Y-7228 below gear change lever cap Y-7220 by tapping slot in ring with screw-driver B-17020. Screw off cap Y-7220 and lift out gear change lever.
Fit dummy cap on gear change housing in place of cap Y-7220 to prevent foreign matter entering gearbox.
Lift off floor board plate YF-940016. 2B

8. Disconnect choke control YE-9700-D from carburettor starting control connection Y-110850 by slacking off rear screw with screw-driver B-17020.

9. Remove oil level indicator YE-6750.

10. Disconnect vacuum wind shield wiper tube YE-17538-B by unscrewing small gland nut B-17542 with wrench Y-810 and bend tube back towards dash.

11. Remove silencer inlet pipe clamp Y-5251 by slacking off two brass nuts, using wrench B-17016 on nuts and wrench B-17015 on bolt heads.

12. Disconnect carburettor to accelerator rod Y-9747 at carburettor. This rod has a spring loaded cap and can be pulled off the ball without difficulty.

Operation   *Illustration*

13. Remove the flexible petrol pipe connection B-9288 at fuel pump with wrench B-17015 and screw special cap on to flexible pipe to stop syphon effect from petrol tank.

14. Remove two screws holding engine front support Y-6030 to front cover Y-6019 with wrench B-17015 noting that off-side screw is longer than near side.

15. Remove four screws which hold clutch housing to cylinder and sump on near-side of clutch housing with wrench B-17015.

16. From underneath car, using creeper 76, remove three nuts and bolts holding near-side engine pan Y-110293-B to frame side member with wrench B-17015. Replace sump drain plug B-6730 with wrench YE-17017 and withdraw drain pan 127 containing oil.

17. Remove screw holding front of near-side engine pan Y-110293-B to front cross member with screw-driver B-17020 and remove engine pan.

18. Remove screw holding timing lever assembly Y-110800 to cylinder head with screw-driver B-17020.

19. Remove four high tension electrical leads, two YE-12275 and two YE-12276 from sparking plugs.
Disconnect high and low tension wires Y-14302 from distributor to coil, at coil B-12000. The high tension wire can be pulled out of socket in coil. The low tension wire is disconnected by taking off nut on near-side of coil with wrench Y-810.

20. Lift off distributor assembly YE-12100 and place in special partition of tray 344.
    CAUTION.—On no account slack off distributor timing lever clamping bolt Y-110807.

21. Remove sparking plugs Y-12405 from cylinder head with box spanner Y-859. Insert dummy plugs 48-M-16 in cylinders Nos. 1, 2 and 4 and eye bolt Y-46 in cylinder No. 3.

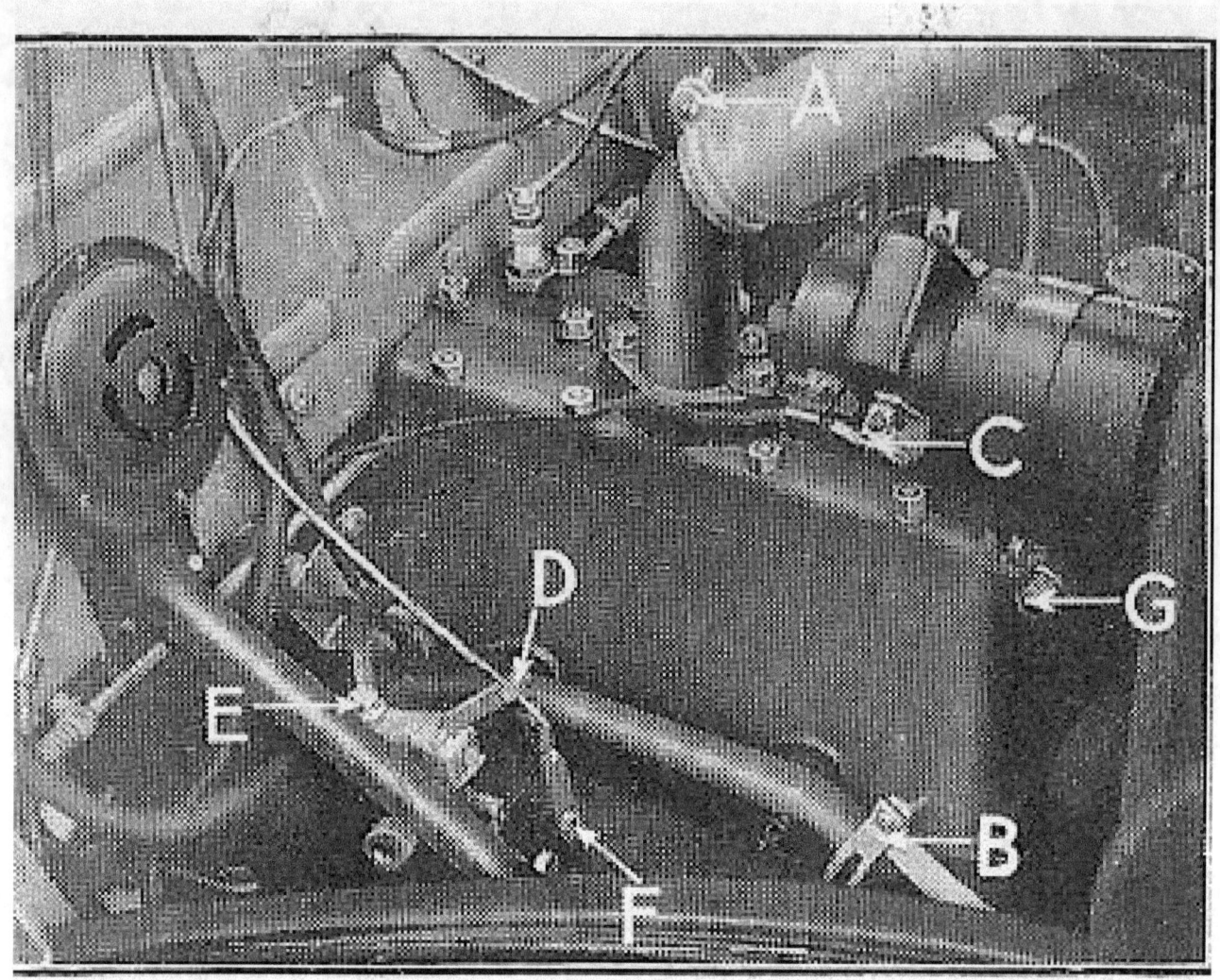

Fig. 3.

Operation   *Illustration*

22. Engage hook of lifting hoist 73 in eye bolt Y-46 and tighten chain until it just takes the weight of the engine.

23. Remove screw holding engine front support Y-6030 to rubber insulator Y-6038 on near-side frame member using wrench 2242 with adapter 2291 and socket 2120.

24. Remove screw holding engine front support Y-6030 to rubber insulator Y-6038 on off-side frame member using wrench 2242 with adapter 2291 and socket 2120 and lift out engine front support Y-6030.

25. Slack off cylinder head outlet hose clamp B-8287 bolts with screw-driver B-17020 work hose B-8260 down on cylinder head outlet connection Y-8250 until it clears radiator inlet, then remove hose from cylinder head outlet connection.   3A

26. Slack off radiator outlet hose rear clamp Y-8287 screw with screw-driver B-17020 and slip end of hose off cylinder inlet water connection Y-8275.   3B

27. Disconnect generator lead by removing screw holding terminal to cut out YE-10505 with screw-driver B-17020 and replace screw.   3C
Remove generator cover band YE-10142 tightening screw, remove clip supporting generator lead and replace screw.

28. Remove screw holding front of off-side engine pan Y-110291-B to front cross member with screw-driver B-17020.

29. Remove starter switch control YE-11475-A cable at starter switch assembly BF-11450 by undoing clamping screw with wrench Y-810.   3D

30. Disconnect battery to switch cable YE-14300-B at switch by unscrewing bolt with wrench 1616.   3E

Operation	Illustration

31. Remove starting motor assembly YE-11000 by unscrewing the two long securing screws YE-11079 with wrench Y-810 the heads of which will be found at the front end of the motor.   3F

32. From underneath car using creeper 76 remove two bolts and nuts holding off-side engine pan Y-110291-B to side member with wrench B-17015 and remove pan.

33. Remove two remaining screws holding clutch housing to cylinder and sump on off-side with wrench B-17015.

34. Close radiator drain cock B-8115 and withdraw drain pan 127 containing water.

35. Slack off nut using wrench 2242, adapter 2291 and socket 2118 on generator support locking pin YE-10159 and tap pin out with copper hammer 83. This will free generator support and allow generator to be lowered. Fan belt YE-8620 can now be worked off generator pulley and over fan blades but should be left on crankshaft pulley Y-6312. Lift off generator YE-10000-B.   3G

36. Raise engine slightly by means of hoist 73 until pulley Y-6312 is clear of front cross member, then move engine slightly forward in order to release transmission main drive gear YE-7015 from clutch disc assembly Y-7550. When clear, lift engine with hoist 73 tilting front end upwards by means of fan belt YE-8620 in order that rear end may clear the dash.

37. When engine is clear of chassis remove fan belt YE-8620 traverse hoist until it is over engine stand AB-35. (If hoist is fixed move chassis back and place engine stand underneath hoist).

38. Run off two nuts which hold cylinder water inlet connection Y-8275 in position with wrench 1616, remove brass washers, inlet connection and gaskets Y-8280. Remove third and fifth sump to cylinder block screws on off-side of engine with same wrench and attach

Operation *Illustration*

38(*cont.*) adapter Y-36 to engine by rear inlet connection stud and Nos. 3 and 5 sump bolt holes using screws already removed.

39. Mount adaptor Y-36 with engine attached to arm of engine stand AB-35 by means of studs provided.

ILLUSTRATIONS AND MEMORANDA

Longitudinal and Cross Sectional Views of the Engine.

TO DISMANTLE ENGINE AND CLUTCH

Special Tools and Equipment Required

Tools from Standard Tool Kit

Wrench ($\frac{7}{16}''$ and $\frac{1}{2}''$)	B-17015
,, ($\frac{9}{16}''$ and $\frac{5}{8}''$)	B-17016
Screw-driver	B-17020
Pliers	B-17025

Special Tools and Equipment previously used

Engine stand	AB-35
Adapter	Y-36
Copper hammer	83
One drain pan	127
Partitioned tray	344
Parts carrier	398
Wrench	1616
Socket	2120
Wrench	2242

Special Tools and Equipment not previously used

Set (2) main bearing wrenches	Y-7
Starting ratchet wrench	Y-31
Piston ring squeezer	Y-38
Pump drive dowel extractor	Y-102
Valve guide remover	Y-113
Valve spring compressor and seat extractor	ABY-322-A & B
Clutch compressor and locator	Y-353
Carburettor wrench	Y-853
Socket	2118
Oil release valve nut wrench	AF-17043

Fig. 5.

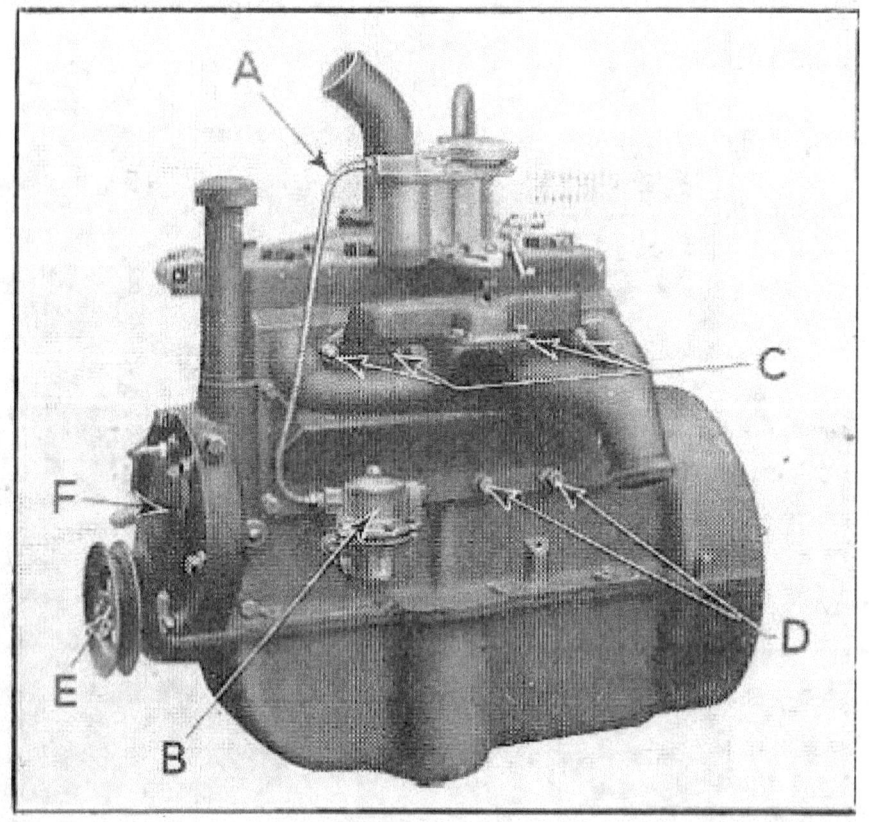

Fig. 6.

IB

TO DISMANTLE ENGINE AND CLUTCH

Carry out in sequence the following operations:—

Operation | Illustration

1. Place drain pan 127 at foot of stand AB-35 underneath engine to catch oil and water drippings.
   Use pan 127 previously employed for drawing water from radiator.

2. Turn engine sump uppermost on stand AB-35 and lock arm in position, when distributor coupling shaft Y-12249 will fall from the hole in the cylinder head and should be placed in tray 344. — 5

3. Remove all remaining engine sump screws using wrench 1616 and remove sump and gaskets. — 5

4. Take out two screws which hold oil pump Y-6600 to cylinder block with wrench Y-853 and remove oil pump.

5. Turn engine back to normal position on stand AB-35 making sure that oil drippings fall into pan 127.

6. Remove petrol pipe YE-9369-D from carburettor and fuel pump using wrench Y-853. — 6A

7. Remove fuel pump YE-9350 and gasket Y-9374 using wrench Y-853. — 6B

8. Remove exhaust manifold stud nuts and washers using wrench 1616, lift off inlet and exhaust manifold assembly with carburettor attached and remove manifold gasket Y-9448. — 6C

9. Remove valve chamber cover Y-6520 and gaskets Y-6521 using wrench 1616, note that top centre bolt is shorter than the remainder to avoid bolt fouling distributor drive shaft. — 6D

10. Remove starting crank ratchet Y-6319 using wrench Y-31 and copper hammer 83. Pull off crankshaft pulley Y-6312. — 6E

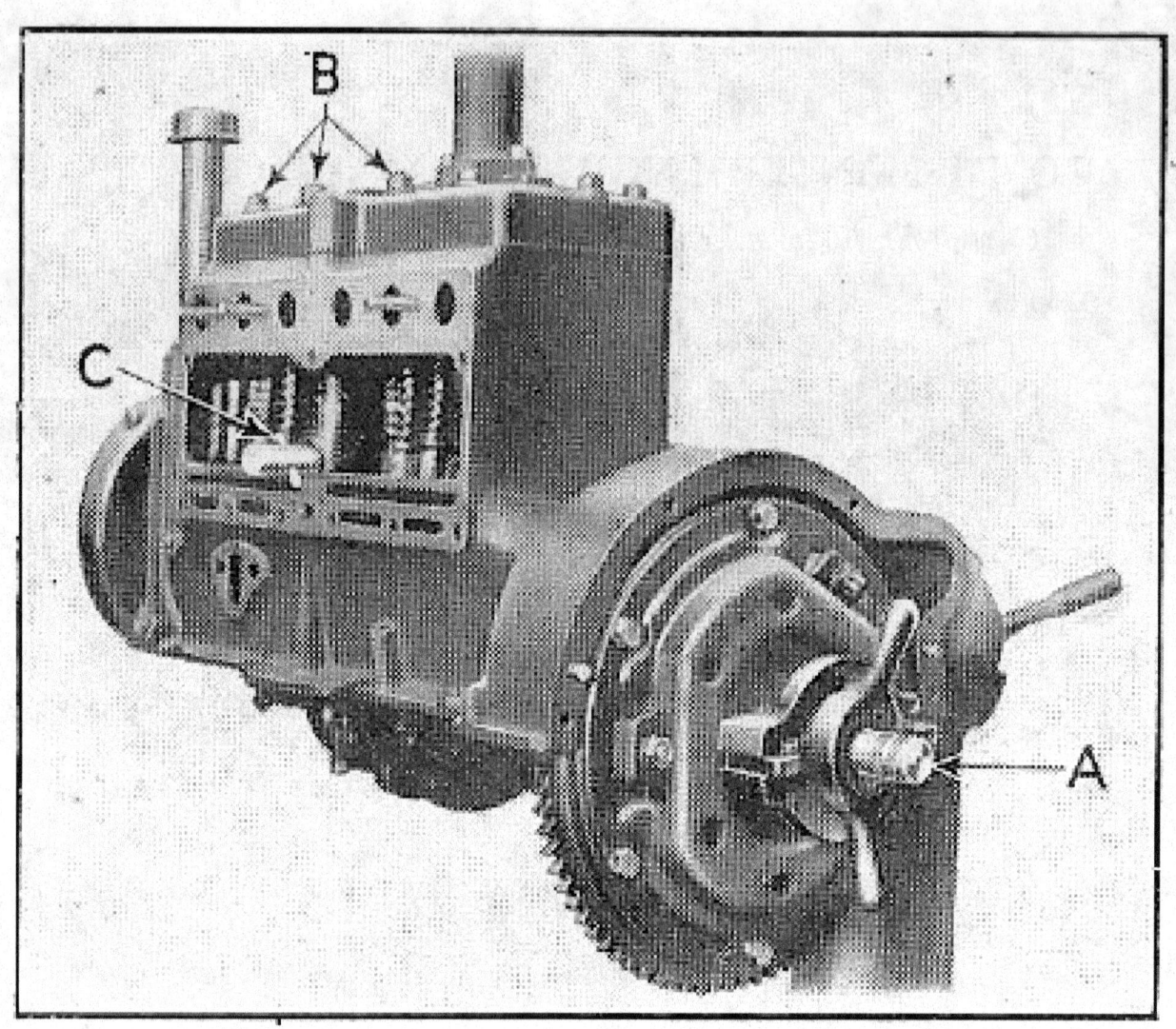

Fig. 7.

1B

Operation		Illustration
11.	Remove cylinder front cover Y-6019 and gasket Y-6020 using wrench 1616 noting that the two bottom screws are longer than the other three. Remove camshaft thrust plunger B-6275 and spring B-6276 from camshaft gear hub and oil slinger Y-6310 from crankshaft. Replace starting crank ratchet Y-6319 by hand.	6F
12.	Remove clutch pressure plate YE-7563 and disc Y-7550 using compressor Y-353 and wrench 1616.	7A

> *NOTE.—When withdrawing clutch pressure plate, clutch disc will fall to the ground unless care is taken to catch this as pressure plate is pulled away from flywheel.*
> *Note that tool Y-353 may be left in position in clutch pressure plate for reassembly.*

13.	Remove cylinder head nuts using wrench 2242 and socket 2118.	7B
14.	Lift cylinder head Y-6050 making sure that head is kept square with studs and remove. Remove cylinder head gasket Y-6051 carefully noting that the turned over edges of gasket are uppermost.	

> *NOTE.—Under no circumstances should the cylinder head be prised up with a screw-driver as this may damage the machined faces of head or block.*

15.	Remove pump drive dowel Y-6566 using tool Y-102.	7C
16.	Remove oil pump drive gear and bearing assembly by lifting out.	
17.	Remove valves as follows :—	

Starting with No. 1 exhaust located at front of valve chamber compress valve spring Y-6513 with compressor ABY-322-A.

Remove valve spring seats Y-6514 with extractor ABY-322-B.

Remove valve springs by keeping valves Y-6505 well off valve seats and easing valve springs Y-6513 outwards from bottom of valve stems.

Remove valve guide bushings Y-6510 with extractor Y-113.

Operation   *Illustration*

17 (*cont.*) Remove valves. This operation is performed automatically when removing extractor Y-113.

> NOTE.—*The valves cannot be withdrawn until the guides have been removed, due to mushroom shape of the valve stems. Care should be taken to ensure that—*
>
> | | |
> |---|---|
> | Valves | Y-6505 |
> | Split guides | Y-6510 |
> | Valve springs | Y-6513 |
> | Spring retainers | Y-6514 |
> | Push rods | Y-6500 |
>
> *are kept in their proper order when placed on parts carrier 398.*

18. Remove push rods Y-6500 by lifting out with the fingers, at the same time turning crankshaft with wrench Y-31 on starting ratchet Y-6319 to lift each push rod.

19. Draw out camshaft and gear assembly Y-6251.

20. Remove split pins only from two centre main bearing bolt nuts with pliers B-17025, one of these is located in the centre of the valve chamber and the other on the opposite side of the cylinder block.

21. Turn engine completely over on stand AB-35, i.e., crankshaft uppermost.

22. Remove split pins from all connecting rod nuts and front and rear main bearing cap nuts using wrench Y-31 on crankshaft ratchet Y-6319 to turn crankshaft to facilitate removal of split pins.

23. Remove No. 1 connecting rod cap nuts using wrench 1616 and lift off cap using copper hammer 83 to ease if necessary.

    Take special note on piece of paper of number of shims YE-6220 (if any) that are used on each side of cap. Rod and cap are stamped with a number on one side which will serve as guide when replacing the shims.

    To withdraw connecting rod and piston assembly take hold of connecting rod with one hand, with the other hand carefully rotate crankshaft by means of wrench Y-31 on starting ratchet Y-6319.

31

29

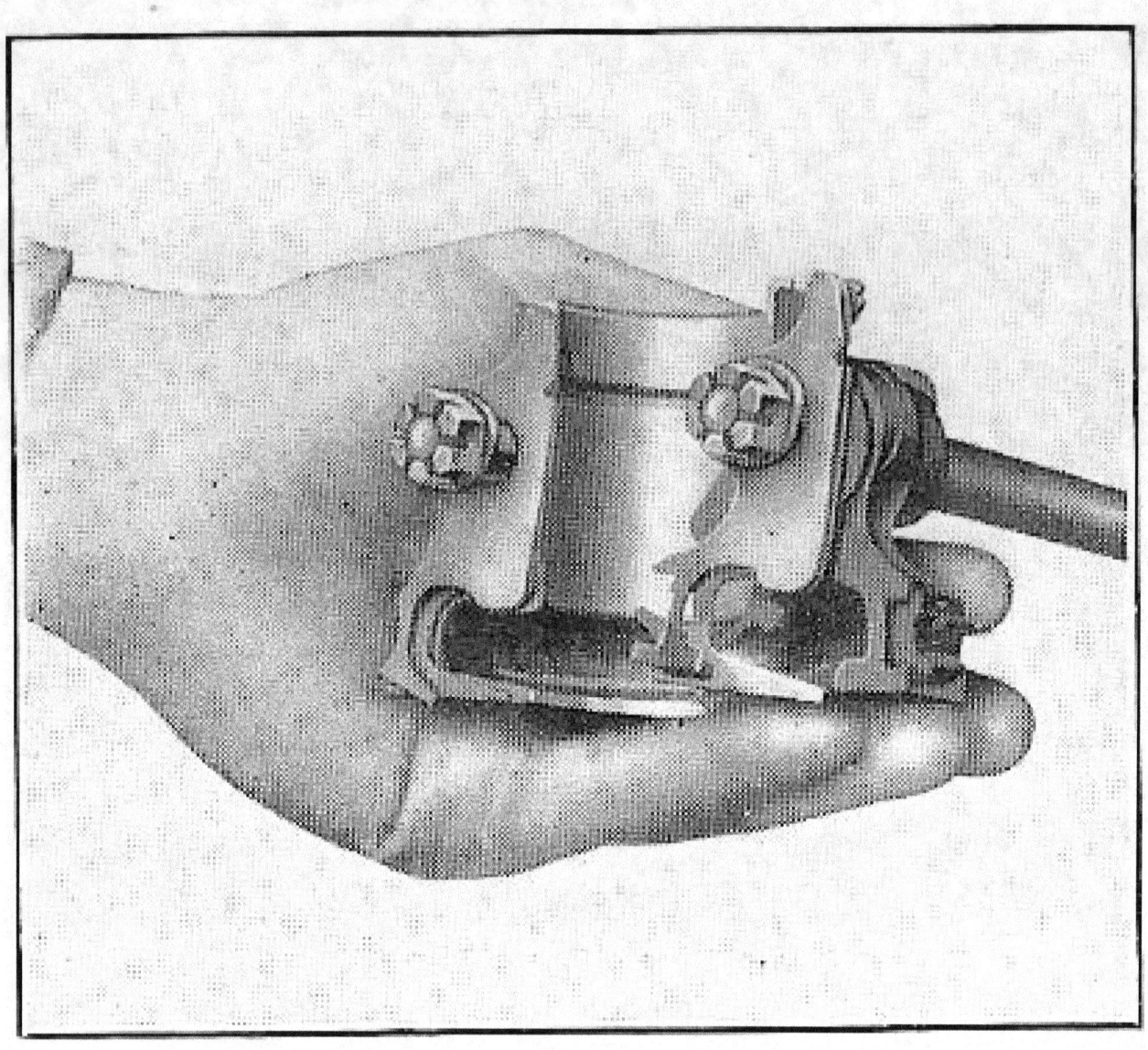

Fig. 8.

Operation	Illustration

23(*cont.*) Connecting rod and piston assembly can be manœuvred and withdrawn past crank webs and counterbalance weights on crankshaft. While doing this care should be taken not to allow piston to drop down in cylinder bore to such an extent that rings come free of bore.

Should this happen it will be necessary to use piston ring squeezer Y-38 to get piston back into cylinder. When connecting rod and piston have been withdrawn replace cap on studs so that numbers on cap and rod correspond.

24. Remove number two, three and four connecting rod and piston assemblies as described in operation number 23.

25. Loosen nuts on main bearing cap bolts using wrenches Y-7. Run nuts off with wrench 2242 and socket 2120 holding other end with wrench Y-7. If bolts will not push out easily they may be tapped lightly with copper hammer 83.

Remove caps, carefully noting number of shims (if any) under each side of each bearing cap. Before placing caps on parts carrier 398, pass bearing bolts through holes, thread correct number of shims over bolt and assemble nut.   8

NOTE.—Special note should be made of brass washer under nut of centre main bearing bolt which is of special finish and is located in the middle of the offside of the cylinder block. This washer and bolt functions as an oil seal.

26. Remove crankshaft Y-6303 complete with flywheel Y-6375 and place on parts carrier 398, flywheel forming base with centre line of crankshaft vertically upright.

Under no circumstances should flywheel and crankshaft be stored other than in manner described when taken out of engine.

27. Turn engine back to normal position on engine stand AB-35 and lock arm in position.

1B

Operation   *Illustration*

28. Remove oil relief valve nut YE-6666-B located at front of engine and adjacent to oil filler pipe Y-6763 using wrench AF-17043, and draw out spring YE-6654-B and valve YE-6663-B.

TO RE-ASSEMBLE ENGINE AND CLUTCH

Special Tools and Equipment Required

Tools from Standard Tool Kit

Wrench	YE-17017
Screw-driver	B-17020
Pliers	B-17025

Special Tools and Equipment previously used

Set (2) main bearing wrenches	Y-7
Starting ratchet wrench	Y-31
Engine stand	AB-35
Adapter	Y-36
Copper hammer	83
Valve spring compressor and spring seat extractor	ABY-322-A & B
Partitioned tray	344
Clutch compressor and locator	Y-353
Parts carrier	398
Carburettor wrench	Y-853
Wrench	1616
Socket	2118
Socket	2120
Wrench	2242
Oil release valve nut wrench	AF-17043

Special Tools and Equipment not previously used

Piston ring squeezer	5M-265-D-1
Oil can	17

37

1C

TO RE-ASSEMBLE ENGINE AND CLUTCH

Carry out in sequence the following operations :—

> NOTE.—With operations marked * in every case before assembling the parts, all bearing surfaces should be wiped clean with a chamois leather and a liberal film of clean engine oil applied with oil can 17.

Operation   *Illustration*

1. Turn cylinder block Y-6010, on engine stand AB-35, so that the main bearings are uppermost and lock arm in position.

2*. Replace the crankshaft and flywheel assembly (Y-6303 and Y-6375) in bearings.

3*. Replace main bearing shims, (see operation 25, section 1B) caps, bolts and nuts starting with centre cap Y-6331, then the rear cap Y-6325, finishing with front cap, Y-6330. Run nuts on with wrench 2242 and socket 2120, finally tightening with wrenches Y-7.

> Note importance of replacing brass washer under nut of centre main bearing bolt of special finish on off-side of engine. (See note to operation 25, Section 1B.)

4. Split pin nuts in front and rear main bearing caps—fit new split pins 72016-S, using pliers B-17025 to open up and bend over legs of pins.

5*. Replace No. 1 connecting rod and piston assembly guiding piston carefully into bottom of cylinder bore. The piston ring squeezer 5M-265-D1, should be fitted over the piston rings before commencing this operation. The operation is a reversal of the dismantling operations described in operation 23, section 1B, and particular attention should be paid to the caution not to allow the piston to drop through the bore for the reason there given.

> Note carefully that the pistons should be replaced so that the split in the skirt is towards the valve chamber, as this is the side of least thrust. Remove piston ring squeezer 5M-265-D1, now released and lying loose on connecting rod.

39

Operation   *Illustration*

6*.   Replace connecting rod cap, shims and nuts on connecting rod assembly, using wrench 1616, tighten up nuts and fit new split pins 72015-S, using pliers B-17025.

7*.   Repeat the operation for remaining connecting rod and piston assemblies.

8.   Remove starting crankshaft ratchet Y-6319, with wrench Y-31, turn engine on engine stand AB-35 to normal position and lock arm.

9*.   Replace camshaft and gear assembly Y-6251, particular care being taken to ensure that the zero or timing mark (a line indicating the space between two teeth) registers with the small punch mark on one of teeth of crankshaft gear Y-6306.

10*.   Replace camshaft thrust plunger B-6275 and spring B-6276 in camshaft gear hub. Replace oil slinger Y-6310 on crankshaft with concave side towards front of engine and locate on crankshaft gear key.

11.   Replace cylinder front cover Y-6019, with gasket Y-6020 and packing Y-6700 and run screws up tight, using wrench 1616. Note that two long screws go in bottom holes.

12.   Replace crankshaft pulley Y-6312, making sure that crankshaft key locates in slot in hub. Replace crankshaft ratchet Y-6319, and tighten with wrench Y-31.

13.   Assemble new split pins 72016-S in centre main bearing bolts using screw-driver B-17020 and pliers B-17025.

14*.   Replace push rods Y-6500 in correct order, tapered end uppermost.

41

Oil passage to
relief valve
Oil passage to
front camshaft
bearing
Latest type gasket. Bevelled corner
Latest type cover. Bevelled
machined face

A
C
F
D
E
B

FIG. 6.

B
C
A

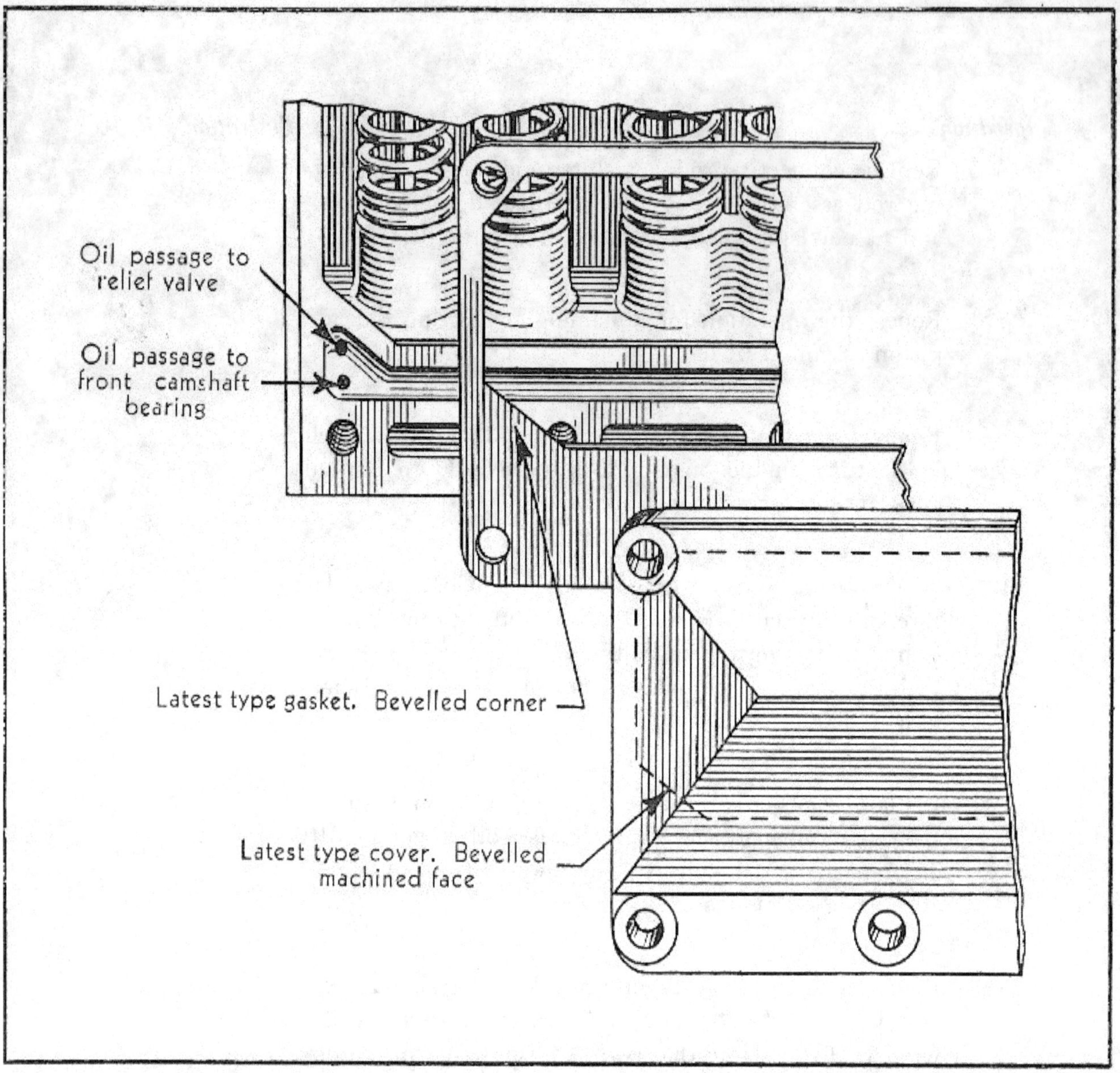

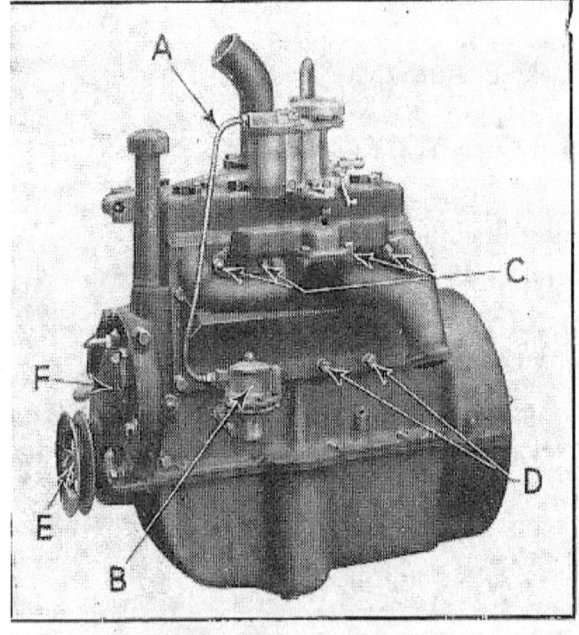

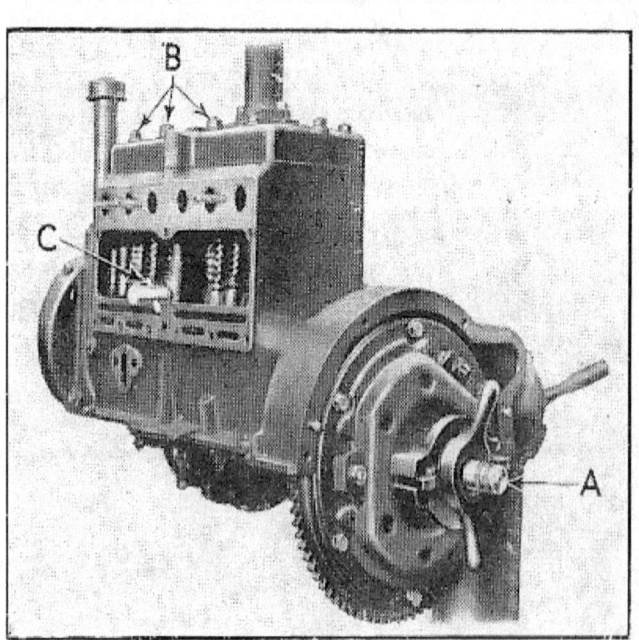

FIG. 7.

Operation *Illustration*

1C

15. Replace valves Y-6505 and valve guide bushings Y-6510, inserting valves from top and split valve guide bushings upwards from valve chamber. It is important to replace valves and valve guide bushings in their correct valve port locations.

16. Replace valve springs Y-6513 by raising valves as high as valve guides will permit. Valve springs can then be inserted easily over valve stems and housed securely against lower portion of valve guide bushings.

17. Replace valve spring retainer Y-6514 by compressing valve spring with compressor tool ABY-322-A and inserting spring retainers with tool ABY-322-B.

18. Remove timing pin YE-6023 using wrench Y-810 from cylinder front cover Y-6019 and insert plain end of pin into hole from which it has been unscrewed. Rotate engine slowly by means of starting crank Y-17036, at same time pressing pin lightly in until pin head drops into indentation in face of bakelized fabric camshaft timing gear Y-6256.

    Replace oil pump drive gear and bearing assembly Y-6551 and Y-6560, so that slot in oil pump drive gear, when viewed from above, lies at an angle of approximately 45° to cylinder front cover Y-6019 with narrower shoulder to rear of engine. Fit dowel Y-6566 making sure that it is located correctly, i.e., fitting flush with cylinder block. Replace timing pin YE-6023 and tighten, using wrench Y-810.

19. Replace valve chamber cover Y-6520, gasket Y-6521, and screws using wrench 1616 ensuring that shorter length bolt is fitted in the centre top location.

20. Replace cylinder head Y-6050 and gasket Y-6051, running nuts down with wrench 2242 and socket 2118. Care should be taken to replace the gasket the right way up (see operation 14, section 1B).

Operation		Illustration
20(*cont.*)	Finally tighten nuts with wrench YE-17017 in the order shown below.	

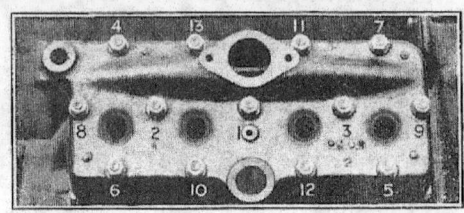

21.	Replace carburettor and inlet and exhaust manifold assembly with gasket Y-9448 clamping in position with washers Y-9443 and nuts using wrench 1616.	6C
22.	Replace fuel pump assembly YE-9350 with gasket Y-9374 using wrench Y-853.	6B
23.	Replace petrol pipe YE-9369-D at fuel pump and carburettor using wrench Y-853.	6A
24.	Replace clutch disc Y-7550 and pressure plate assembly YE-7563 using special compression tool Y-353 and wrench 1616.	7A

NOTE.—Care should be taken to ensure that the pressure plate housing is located correctly in register on flywheel before tightening screws.

25.	Remove compressor and location tool Y-353.
26.	Turn engine over on engine stand AB-35 so that crankshaft is uppermost and lock arm in position.
27.	Replace oil pump assembly Y-6600 and tighten screws using wrench Y-853.
28.	Replace sump Y-6675 with gaskets Y-6701-6710-6711, insert and tighten bolts leaving out the three bolts on off-side of sump where the adapter Y-36 is clamped to the cylinder block.

1C

Operation   *Illustration*

29. Turn engine over on engine stand AB-35 and lock arm in position.

30. Replace distributor coupling shaft Y-12249 making sure that tongue on shaft engages in slot in oil pump drive gear shaft using screw-driver B-17020 to turn coupling shaft.

31. Assemble oil relief valve plunger YE-6663-B to spring YE-6654-B and insert in cylinder block. Replace nut YE-6666-B making sure that stem enters spring YE-6654-B and that copper asbestos gasket Y-6653 is in position under nut. Tighten nut down with wrench AF-17043.

> *IMPORTANT.—If it is desired to remove and dismantle the gearbox, the operations covered in Section 2 should be carried out at this point before proceeding with Section 1D.*
>
> *It is not possible to remove and replace the gearbox without first taking out either the engine or the rear axle, and it is desirable, therefore, to deal with the gearbox operation at the present stage.*

TO INSTALL ENGINE AND CLUTCH IN CHASSIS

Special Tools and Equipment Required

Tools from Standard Tool Kit

Double ended wrench $\frac{7}{16}''$ and $\frac{1}{2}''$	B-17015
,, ,, ,, $\frac{9}{16}''$ and $\frac{5}{8}''$	B-17016
Screw-driver	B-17020
Pliers	B-17025
Starting crank	YE-17036
Jack assembly	YE-17080

Special Tools and Equipment previously used

Engine stand	AB-35
Engine stand adapter	Y-36
Eye bolt	Y-46
Lifting hoist	73
Partition tray	344
Parts carrier	398
Wrench	Y-810
,,	Y-853
Sparking plug spanner	Y-859
Wrench	1616
Socket	2118
,,	2120
Wrench	2242
Universal adapter	2291

Special Tools and Equipment not previously used

Socket	2116
Bar handle	2256
Short " T " wrench	2263
Extension bar	2298

TO INSTALL ENGINE AND CLUTCH IN CHASSIS

Carry out in sequence the following operations:—

Operation   Illustration

1. Attach hook of lifting hoist 73 to eye bolt Y-46 and remove engine and clutch, with adapter Y-36 attached, from stand AB-35. With engine and clutch suspended from hoist 73 remove adapter Y-36 and replace three sump screws using wrench 1616. Replace cylinder water inlet connection Y-8275 with gaskets Y-8280 brass washers 34756-S and nuts using wrench 1616. Replace fan belt over crankshaft pulley Y-6312.

2. Move engine and clutch over chassis and lower gently with hoist 73, at the same time tilting front end upwards by means of fan belt YE-8620 in order that rear end may clear the dash, and when clear lower front end until crankshaft pulley Y-6312 is just over but not resting on front cross member Y-5020, then move engine and clutch back in order to engage the splined transmission main drive gear Y-7015 with splined clutch disc assembly Y-7550.

3. Replace top screw in clutch housing loosely. Replace bottom screw in clutch housing using creeper 76 and remove jack assembly YE-17080 which was used when installing gearbox.

4. Replace front engine support Y-6030 to rubber insulator Y-6038 and replace both screws using wrench 2242 with adapter 2291 and socket 2120.

   NOTE.—When installing support Y-6030, care should be taken not to trap the fan belt YE-8820 between the support and the front of the engine.

5. Lower engine and clutch on to front engine support Y-6030 at the same time guiding the radiator outlet hose Y-8286 on to cylinder water inlet connection Y-8275.

Fig. 4.

Operation *Illustration*

5(*cont.*) Rear clamp Y-8287 should be loose on hose. Replace and tighten two screws holding engine front support Y-6030 cylinder front cover Y-6019 using wrench B-17105 and noting that the off-side screw is longer than the near-side screw.

6. Remove hoist 73. Remove eye bolt Y-46 and dummy plugs 48-M-16 from cylinder head and replace sparking plugs Y-12405 using box spanner Y-859.

7. Remove special cap from flexible petrol pipe connection B-9288 and replace pipe at fuel pump using wrench B-17015. Replace carburettor to accelerator rod Y-9747 at carburettor. This rod has spring loaded cap and can be pushed on ball without difficulty.

8. Replace silencer inlet pipe clamp Y-5251 noting that the narrow flange face is uppermost, using wrenches B-17015 and B-17016.

9. Screw out timing pin YE-6023 using wrench Y-810 from cylinder front cover Y-6019 and insert plain end of pin into hole from which it has been unscrewed. Rotate engine slowly by means of starting crank Y-17036, at same time pressing pin lightly until pin head drops into indentation in face of timing gear Y-6256. Take off distributor cap YE-12116-B and make sure that rotor contact is facing No. 1 cylinder contact position. Place distributor YE-12100-B in engine and secure body clamp plate to cylinder head by means of screw 26475-S, plain washer Y-110819, and lock washer 34805-S with zero reading of scale against index mark on cylinder head. Loosen body clamp screw 26475-S and move body clamp until " 4 " graduation mark of scale is against index mark. Finally tighten down body clamp screw 26475-S.

10. Attach choke control YE-9700-D to carburettor starting   4A
control connection Y-110850 using screw-driver B-17020.
The inner cable should be connected so that choke button

Fan Belt Adjustment.

1D

Operation *Illustration*

10(*cont*.) on dash has approximately $\frac{1}{64}$" "lost motion." This is to prevent movement of engine causing rich mixture when choke is closed.

11. Replace vacuum wind shield wiper tube YE-7538-B by screwing up small gland nut B-17542 using wrench Y-810.   4B

12. From underneath car using creeper 76 offer up near-side engine pan Y-110293-B. From above, replace screw to hold near-side engine pan to front cross member Y-5020 using screw-driver B-17020.

13. From underneath car using creeper 76 replace three bolts, spring washers and nuts to hold near-side engine pan to frame side member using wrenches B-17015 and Y-853. For ease of operation assemble all bolts loosely starting from the rear.

14. Replace off-side engine pan Y-110291-B proceeding in same manner as for operation 13.

15. Replace starting motor assembly YE-11000 by screwing up two long securing screws YE-11079-B using wrench Y-810.

16. Replace starter switch control YE-11475-A cable at starter switch assembly BF-11450 and tighten up clamping screw with wrench Y-810. Replace battery to switch cable YE-14300-B and lead from main wiring assembly to starter motor switch BF-11450 with screw using wrench 1616.

   NOTE.—The main wire assembly lead terminal should be located under the battery to switch cable terminal.

17. Replace generator and cut-out assembly YE-10000-B and work fan belt YE-8620 into position over generator pulley. Replace generator support locking pin YE-10159 from the near-side and replace spring washer and nut, raise generator assembly sufficient to allow one inch slack in fan belt YE-8620 and tighten up nut using wrench 2242, adapter 2291 and socket 2118.

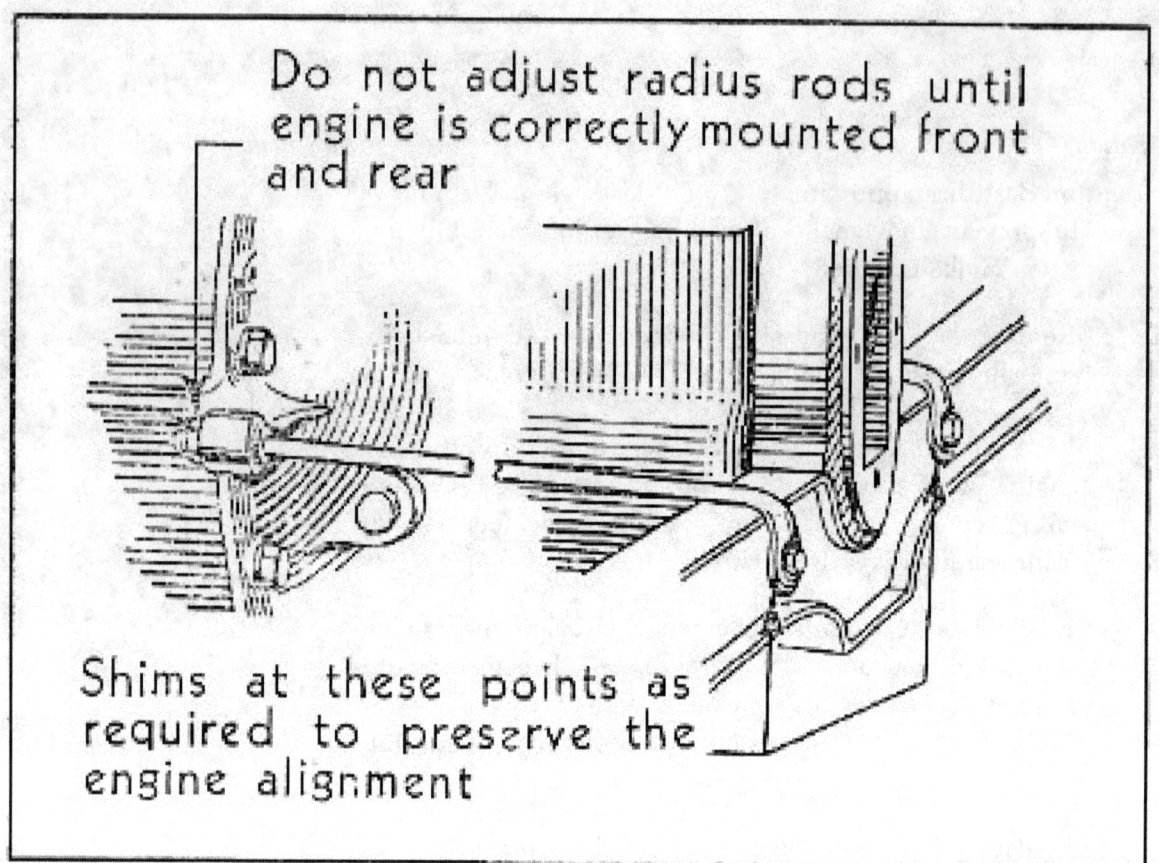

Operation	Illustration

18. Remove generator cover band tightening screw, replace clip supporting generator lead and replace screw. Remove terminal screw on off-side of cutout YE-10505, replace generator lead and cutout terminal screw, using screw-driver B-17020. Move radiator outlet hose rear clamp Y-8287 into position and tighten screw.

19. Replace cylinder head outlet hose Y-8260 and clamp B-8287 and tighten up clamp screws using screw-driver B-17020.

20. Replace remaining clutch housing screws and tighten up, using wrenches 1616, B-17015 and creeper 76.

21. Check engine radius rod Y-6028 for possible clearance between turned over rear end of rod and chassis cross member YR-5025. If there is clearance remove screws and use shims YE-6054-A or B to rectify clearance, in order to avoid misaligning engine when tightening up screws. Replace screws and washers and tighten with wrench B-17015.

22. Replace floor board plate YF-940016. Replace two screws holding front of floor board plate to dash. Replace three screws holding rear of floor board plate to cross member. Replace foot dimmer switch Y-110378 to floor board plate YF-940016 and replace screws using screw-driver B-17020 for all these screws.

23. Replace No. 2 (horizontal) off-side floor board YF-940132-B in position and screw in the three fixing screws. Replace No. 2 near-side floor board YF-940133-B in position and screw in the three fixing screws. Replace No. 1 off-side floor board YF-940130-B in position and screw in the eight fixing screws noting that the two screws at the top of this floor board are shorter than the rest. Replace No. 1 near-side floor board YF-940131-B in position and screw in the six fixing screws.
Screw-driver B-17020 should be used for the fixing screws of the above floor boards.

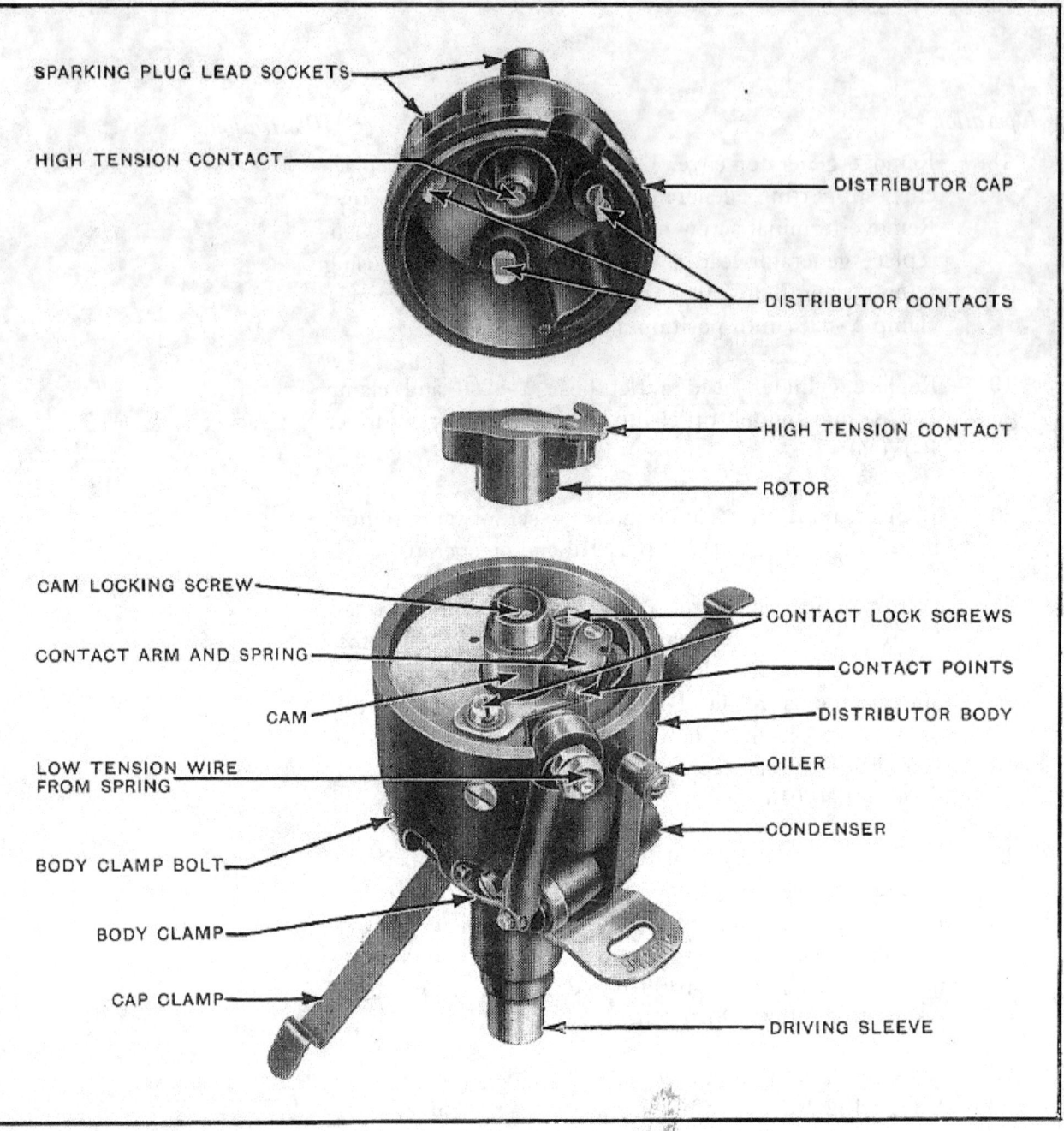

ID

Operation *Illustration*

24. Remove dummy cap on gear change housing and replace gear change lever assembly screwing home cap Y-7220 by hand and tapping locking ring Y-7228 up against cap Y-7220 with screw-driver B-17020.

25. Replace accelerator pedal pad YE-11471 screwing pad up by hand and locking up nut against pad using wrench B-17015.

I.D. 26. Reset ignition timing proceeding in following manner. Make sure that timing pin YE-6023 is still engaged in indentation of timing gear Y-6256. Slacken off distributor body clamp screw 26475-S. Turn distributor body in clockwise direction until contact breaker points are just about to open. This should occur while condenser YE-12300-B is facing cylinder head. Tighten down body clamp screw 26475-S. Replace rotor YE-12200-B on centre shaft engaging tongue in slot on cam. Replace cap YE-12116-B and secure by holding down clips. Connect four outer H.T. leads from cap to sparking plug terminals (noting that two leads YE-12275 to Nos. 2 and 3 plugs are shorter than leads YE-12276 to Nos. 1 and 4 plugs) and centre H.T. lead (in loom Y-14302) to the coil in the following order :

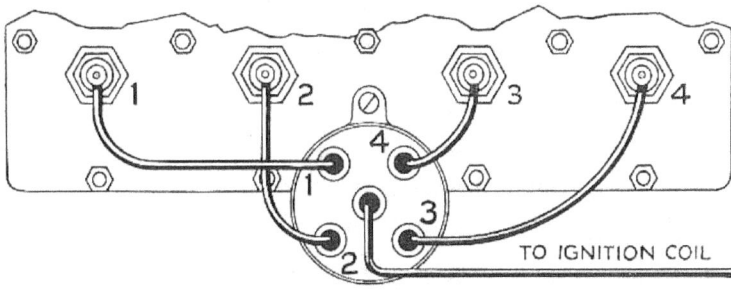

Replace L.T. lead (in loom Y-14302) from distributor to coil B-12000 using wrench Y-810. Replace timing pin Y-6023 correct way round in cylinder front cover and tighten up using wrench Y-810.

Operation *Illustration*

27. Replace battery negative terminal lead by removing the two cover fastening nuts Y-110932 with screw-driver B-17020 and lifting cover clamping lead to negative terminal post using wrench B-17015. Coat battery terminals liberally with vaseline to prevent corrosion. Replace battery cover and tighten up fastening nuts.

28. Check radiator water drain cock to assure that it is in closed position. Remove cap Y-8109 fill radiator with water and replace cap. Check all water joints for possible leaks and if any, tighten up connections at fault.

29. Remove oil filler cap Y-6766 and pour into oil filler pipe Y-6763 slowly one half gallon of fresh engine oil, replace filler cap Y-6766. Replace oil level indicator YE-6750.

30. Replace the bonnet in the following manner. Pick up bonnet and offer it up to cowl and radiator shell in such a position that when bonnet rests on car the side nearest operator is closed leaving other side folded back. It does not matter from which side bonnet is offered up. Guide front end of bonnet hinge into hinge bracket YE-8248 on radiator and rest bonnet in position on cowl at rear end. Replace rear bonnet hinge bracket Y-16729 and drop screw into position. Close open side of bonnet and raise other side. Offer radiator brace rod Y-8132-B into position at front and rear and replace other rear hinge bracket screw. This will now pass through rear end of brace rod, run nut on to screw and tighten up using wrench B-17015 and screw-driver B-17020. Close open side of bonnet and raise other side. Run nut on to screw in rear hinge bracket and tighten up. Finally adjust bonnet setting by screwing adjusting nuts at front end of brace rod Y-8132-B left or right as the case may be.

31. Before starting up engine make quite sure that gear lever is in neutral position.

Section 2

GEARBOX

A. To remove gearbox from chassis.

B. To dismantle gearbox.

C. To re-assemble gearbox.

D. To install gearbox into chassis.

TO REMOVE GEARBOX FROM CHASSIS

Special Tools and Equipment Required

Tools from Standard Tool Kit

Wrench $\frac{7}{16}''$ and $\frac{1}{2}''$	B-17015
,, $\frac{9}{16}''$ and $\frac{5}{8}''$	B-17016
,, adjustable	B-17021
Pliers	B-17025
Jack assembly	YE-17080

Special Tools and Equipment previously used

Engine stand	AB-35
Creeper	76
Copper hammer	83
Drain pan	127
Partition tray	344
Parts carrier	398
Socket	2118
,,	2120
Wrench	2242
Adapter	2291
Bar handle	2256
"T" wrench	2263
Extension	2298

Special Tools and Equipment not previously used

Engine stand adapter	Y-416

Fig. 9.

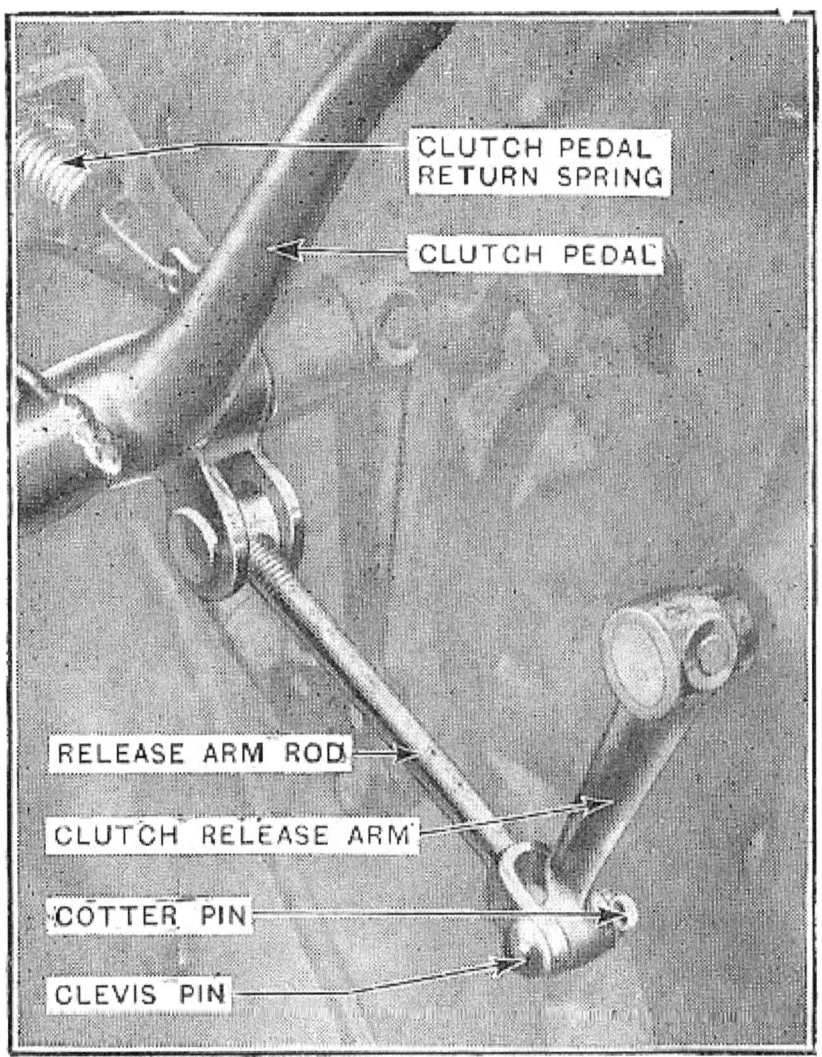

TO REMOVE GEARBOX FROM CHASSIS

Carry out in sequence the following operations:—

Operation		Illustration
1.	Place drain pan 127 under gearbox and remove drain plug Y-24452 using creeper 76 and wrench B-17021.	
2.	Remove locking wire from universal joint housing cap bolt heads using pliers B-17025.	9A
3.	Remove split pins from the two nuts at the engine rear support strap Y-5103 using pliers B-17025.	9B
4.	Remove split pin and clevis pin from the clutch release shaft arm Y-7511 using pliers B-17025.	9C
5.	At this point we must support gearbox from underneath using jack assembly YE-17080, before proceeding to release nuts and bolts holding it in position.	
6.	Remove two nuts from engine rear support strap Y-5103 using wrench 2242 adapter 2291 and socket 2120.	9B
7.	Remove strap Y-5103. This may need easing off the two bolts passing through ends of strap; if so, a smart tap with copper hammer 83 will generally allow strap to spring off.	9D
8.	Remove three upper bolts from universal joint housing cap using wrench 2263, bar 2256 and socket 2118.	9E
9.	Remove two bolts which are screwed in chassis centre cross member from rear end of engine radius rods Y-6028 using wrench B-17015.	9F
10.	From underneath chassis using creeper 76 remove remaining bolt from universal joint housing cap using wrench 2263, bar 2256, extension 2298 and socket 2118.	

2A

Operation *Illustration*

11. The gearbox is now free in chassis and it should be eased forward so as to clear splines on drive shaft Y-4605-B from splines in universal joint Y-7090 and lifted out of chassis.

12. Place gearbox on engine stand AB-35 by means of adapter Y-416.

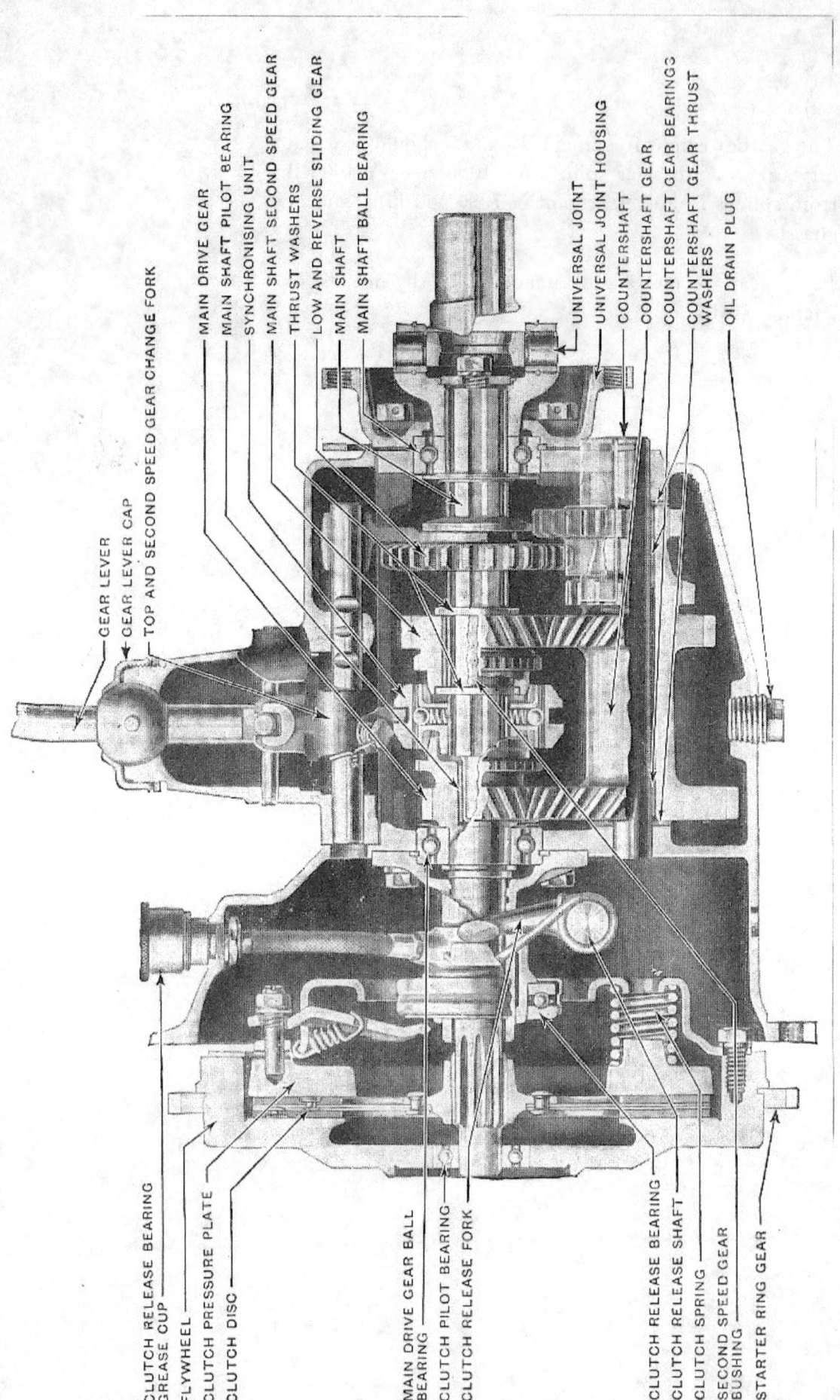

Clutch and Gearbox.

TO DISMANTLE GEARBOX

Special Tools and Equipment Required

Tools from Standard Tool Kit

Wrench $\frac{7}{16}''$ and $\frac{1}{2}''$	B-17015
,, $\frac{9}{16}''$ and $\frac{5}{8}''$	B-17016
Pliers	B-17025

Special Tools and Equipment previously used

Engine stand	AB-35
Copper hammer	83
Partition tray	344
Parts carrier	398
Engine stand adapter	Y-416
Wrench	1616
Socket	2118
Wrench	2242

Special Tools and Equipment not previously used

Snap ring remover	AATA-8
Brass drift	382
Wrench : clutch grease lock nut	Y-811

Fig. 10.

Fig. 11.

2B

TO DISMANTLE GEARBOX

Carry out in sequence the following operations:—

Operation		Illustration
1.	Remove two engine radius rods Y-6028 using pliers B-17025 and wrench B-17016.	
2.	Remove four bolts in gear change housing Y-7222 using wrench 1616,	10A
	SPECIAL NOTE.—At no time when gear change housing has been removed must outer member of synchronising unit be moved in relation to the inner member. This is extremely important.	
3.	First remove clutch release bearing grease connections Y-7557 by unscrewing lock nut with wrench Y-811 then unscrew connection from clutch release bearing Y-7580 using wrench B-17015.	11A
4.	Remove clutch release bearing hub spring Y-7562 from hub Y-7561 using pliers B-17025.	11B
5.	Remove clutch release bearing Y-7580 and hub Y-7561 assembly by sliding forward on transmission main drive shaft.	11C
6.	Remove pin 72257-S which passes through centre of shaft YR-7510 and through centre of clutch release fork Y-7515 using copper hammer 83 to drift out pin.	11D
7.	Remove clutch release bearing shaft YR-7510 by drawing it out of gearbox case from off-side.	11E
8.	Remove three bolts in main drive bearing retainer Y-7050 using wrench 1616.	11F
9.	Remove snap ring Y-7070 from main shaft bearing using tool AATA-8.	
10.	Remove universal joint Y-7090. To carry out this operation use wrench 1616 and unscrew bolt located in rear end of main shaft Y-7061. When this bolt is withdrawn note that retainer Y-7095 is not dropped or mislaid as it can easily be overlooked.	10B

Fig. 12.

2B

Operation   *Illustration*

11. Remove locking wire from four bolt heads at main shaft bearing retainer Y-7085 using pliers B-17025.

12. Remove the four main shaft bearing retainer bolts using wrench 2242 and socket 2118.   10C

13. Remove bearing retainer Y-7085 and retainer Y-7155 now released by operation 12.

14. Remove main shaft ball bearing assembly Y-7065 from main shaft Y-7061 taking care of main shaft oil baffle Y-7080. This bearing should come out quite easily now retainer Y-7085 has been removed.

15. Remove transmission main drive bearing assembly Y-7065 from shaft Y-7015 taking care not to damage baffle Y-7040. This bearing should come out quite easily now retainer Y-7050 and snap ring Y-7070 have been removed.

16. Remove synchronising unit Y-7106 and Y-7108 by pulling two ends of shafts Y-7015 and Y-7061 apart. This action will enable synchronising unit to be drawn off main shaft Y-7061 and so lifted out through top of gearbox.   12

17. Lift main shaft Y-7061 assembly out through top of gearbox.

18. Lift transmission main driving gear and shaft Y-7015 assembly out through top of gearbox.

19. Remove counter-shaft Y-7111 by drifting it out with brass drift 382 and copper hammer 83 from the front end of the gearbox. The retainer Y-7155 which locks this shaft in position was removed in operation 13.

20. Remove counter-shaft gear assembly Y-7114 by lifting out through top of gearbox.

    SPECIAL NOTE.—In carrying out operation 20, careful note should be made of two bronze thrust washers Y-7119 which will have fallen into bottom of gearbox. These washers should be retrieved and placed with small miscellaneous parts in partitioned tray 344.

Operation *Illustration*

21. Remove reverse idler shaft Y-7140 by drifting out with brass drift 382 and copper hammer 83 from inside of gearbox.

22. Remove the reverse idler gear Y-7141 by lifting it out through top of gearbox.

TO RE-ASSEMBLE GEARBOX

Special Tools and Equipment Required

Tools from Standard Tool Kit

Wrench $\frac{7}{16}''$ and $\frac{1}{2}''$	B-17015
,, adjustable	B-17021
Pliers	B-17025

Special Tools and Equipment previously used

Snap ring remover	AATA-8
Engine stand	AB-35
Copper hammer	83
Partition tray	344
Parts carrier	398
Engine stand adapter	Y-416
Wrench clutch greaser lock-nut	Y-811
Wrench	1616
Socket	2118
Wrench	2242

Special Tools and Equipment not previously used

Second gear lock plunger press	BV-3
Main drive gear bearing press	Y-8-A
Main shaft bearing press	Y-8-B
Synchromesh gear ball ring	Y-8-J
Counter shaft gear assembly tool	Y-8-K

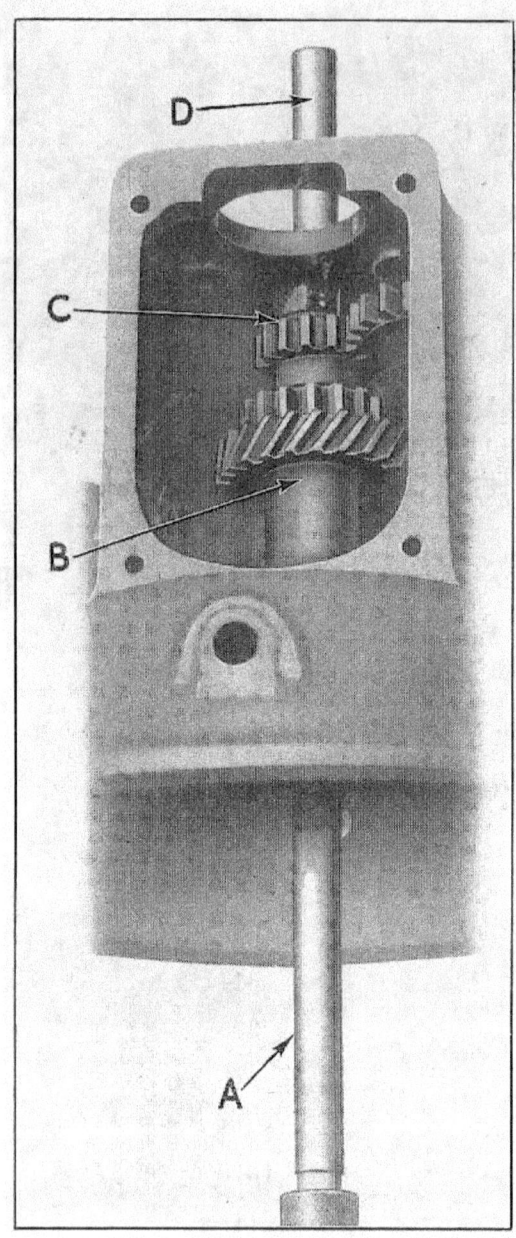

Fig. 13.

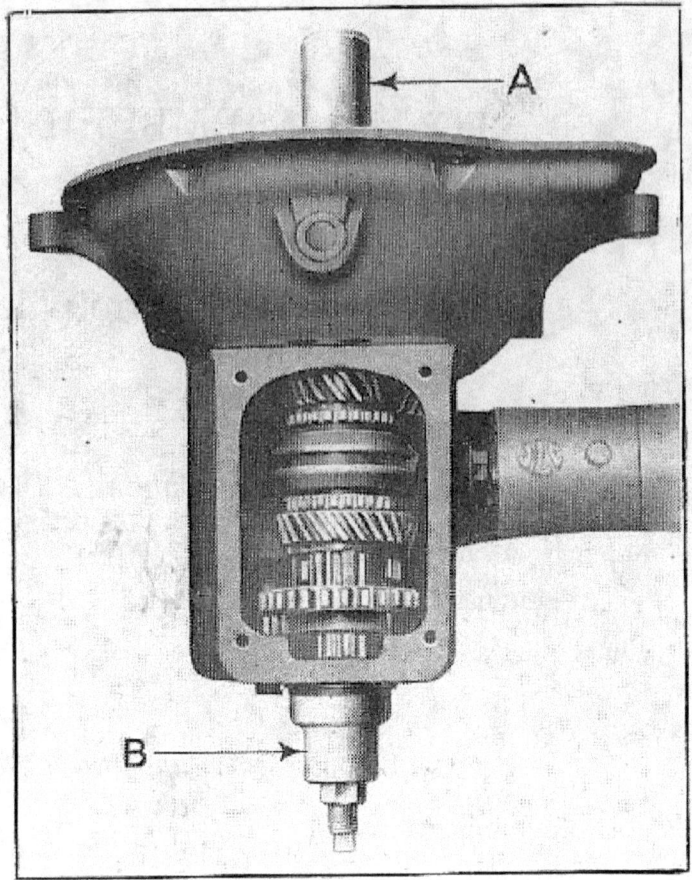

Fig. 14.

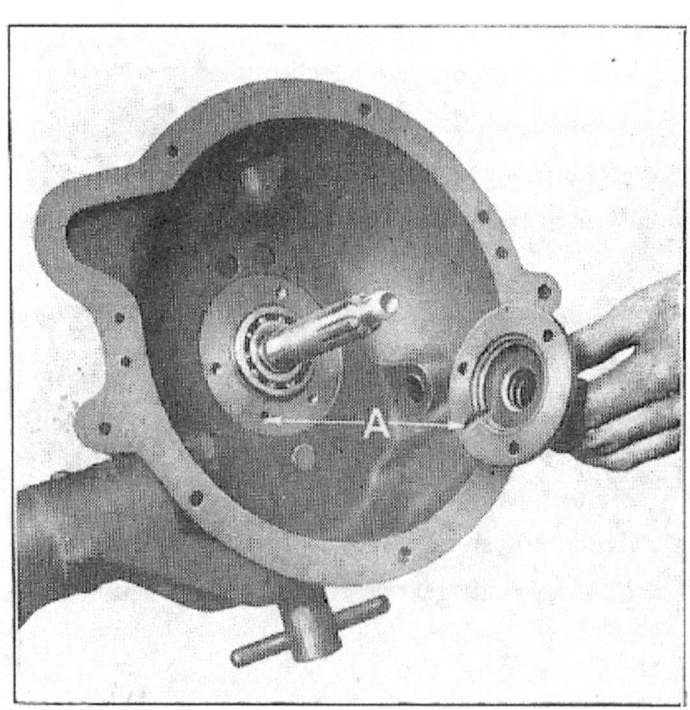

Fig. 15.

2c

TO RE-ASSEMBLE GEARBOX

Carry out in sequence the following operations :—

Operation *Illustration*

1. Replace reverse idler gear Y-7141 in gearbox—hold in position for next operation.

2. Replace reverse idler gear shaft Y-7140 using copper hammer 83 to drive in shaft, taking care to see that it enters into reverse idler gear bushing as in operation 1.

   NOTE.—Shaft should be inserted so that when assembled in gearbox, slot in end of shaft projects from rear face of gearbox.

   Turn gearbox with universal joint end uppermost and lock arm of engine stand in this position.

3. Replace bronze thrust washer Y-7119 at front bushing Y-7121 of counter-shaft Y-7111 inside gearbox.

4. Insert pilot Y-8-K from outside of gearbox and through bushing Y-7121 and locate bronze thrust washer. — 13A

5. Replace counter-shaft gear assembly Y-7114 into gearbox. — 13B

6. Push pilot Y-8-K up to locate in counter-shaft gear assembly Y-7114.

7. Replace bronze thrust washer Y-7119 at rear end of countershaft gear assembly Y-7114. — 13C

8. Push pilot Y-8-K up to locate in bronze thrust washer Y-7119 and also rear bushing of countershaft Y-7111 in gearbox.

9. Replace countershaft Y-7111 from rear end of gearbox, i.e., universal joint end, pushing pilot Y-8-K out as countershaft Y-7111 is inserted using copper hammer 83 if necessary. — 13D

   Turn the gearbox on engine stand back to horizontal and lock arm in position.

Operation	Illustration

10. Replace transmission main drive gear Y-7015 in gearbox.

11. The synchromesh assembly having been previously assembled on main shaft Y-7061—the whole is now replaced in gear box.

12. Replace main-shaft ball bearing Y-7065 noting that thick oil baffle Y-7080 is assembled on inside of bearing. Press tool Y-8-B is used for replacement of bearing which is carried out in following manner :—
Centre screw of press Y-8-B is first screwed into the threaded hole in centre of main shaft Y-7061 then by screwing down the large hexagon nut, using wrench B-17021 for this, the outer sleeve of the press will push the bearing home.   14B

13. Replace transmission main drive ball bearing Y-7065 noting that thin oil baffle Y-7040 is assembled on the inside of the bearing. Use tool Y-8-A for this operation in the following manner :—
Place tool Y-8-A over shaft Y-7015 and guide bearing Y-7065 into line tapping back end of tool Y-8-A with copper hammer 83 to push bearing into place. Remove tool Y-8-A.   14A

14. Replace main shaft retainer snap ring Y-7070 using tool AATA-8 for this operation.

15. Remove press Y-8-B.

16. Replace countershaft Y-7111 and remove idler shaft Y-7140 and retainer Y-7155 making sure that it engages in the slots at the rear ends of the shafts Y-7140 and Y-7111 which project from the rear face of the gearbox.

17. Replace the main shaft bearing retainer Y-7085 and bolts using wrench 2242 and socket 2118 to tighten screws. These should be locked by threading a length of wire through hole in each screw head and locking ends together.

2C

Operation *Illustration*

18. Replace the universal joint Y-7090 on the splines of the main shaft Y-7061.

19. Insert retainer Y-7095 in universal joint, replace screw and tighten up using wrench 1616.

20. Replace gear change housing Y-7222 taking care to see that selector forks Y-7230 and Y-7231 are engaged correctly in grooves provided for them on outer member of synchromesh unit and low and reverse sliding gear. Insert housing screws and tighten them using wrench 1616.

21. Replace main drive gear bearing retainer Y-7050 and bolts using wrench 1616.

    NOTE:—In carrying out this operation it is essential that the oil return slot registers with the hole in the gearbox casing. 15A

22. Replace clutch release shaft YR-7510 from off-side of box but do not push this shaft into its final position—If this is done it will be impossible to assemble clutch release shaft fork Y-7515.

23. Assemble clutch release shaft fork Y-7515 and clutch release bearing spring Y-7562 on clutch release shaft YR-7510 and push shaft home into near-side bearing.

24. Line up holes in clutch release shaft fork Y-7515 and clutch release shaft YR-7510 and insert clutch release shaft fork pin, tapping it home with copper hammer 83.

25. Replace clutch release bearing hub assembly and with pliers B-17025 slip long arm of spring Y-7562 in position on ear of hub Y-7561.

26. Insert clutch release bearing grease connection Y-7557 through the hole in top of clutch housing and screw it into hub Y-7561 first, then tighten up lock nut using wrenches B-17015 and Y-811.

27. Replace two engine radius rods Y-6028 and run nuts on finger tight only.

28. Replace drain plug Y-24452 using wrench B-17021.

TO INSTALL GEARBOX INTO CHASSIS

2D

Special Tools and Equipment Required

Tools from Standard Tool Kit

Wrench $\frac{9}{16}''$ and $\frac{5}{8}''$	B-17016
" adjustable	B-17021
Pliers	B-17025
Jack assembly	YE-17080

Special Tools and Equipment previously used

Engine stand	AB-35
" " adapter	Y-416
Socket	2118
"	2120
Wrench	2242
Bar handle	2256
" T " wrench	2263
Adapter	2291
Extension	2298

Special Tools and Equipment not previously used

Rear engine support strap jig	5M-359

TO INSTALL GEARBOX INTO CHASSIS

Carry out in sequence the following operations :—

Operation   Illustration

1. Remove gearbox from engine stand adapter Y-416 by unscrewing special plug holding gearbox in position.

2. Replace inner universal joint housing cap Y-4513 and gasket Y-4515 and hold them in position.
    NOTE.—Universal joint should be liberally packed with good grade grease.

3. Offer gearbox into position in chassis taking care to engage splines of drive shaft Y-4605-B into corresponding splines of universal joint assembly Y-7090.

4. Replace three upper screws in universal joint housing cap and tighten them up evenly using wrench 2263, bar handle 2256 and socket 2118.

5. From underneath chassis using creeper 76 replace one remaining universal joint housing cap screw and tighten it up using wrench 2263, bar handle 2256, extension 2298 and socket 2118.

6. While still under chassis place jack assembly YE-17080 under gearbox and screw jack up enough to take weight of box.

7. Thread length of wire through holes in four screw heads of universal joint housing cap screws, locking ends of wire together.

8. Replace two screws through rear end of engine radius rods into chassis centre cross member but do not tighten them.

9. Offer into position engine rear support strap Y-5103.

Operation *Illustration*

10. Replace off-side engine rear support strap nut, running it down far enough to obtain good grip of threads, then insert jig 5M-359 to give necessary leverage to enable other nut to be run on. Finally remove jig 5M-359 tighten two nuts down evenly and split pin them using wrench 2242, adapter 2291, socket 2120 and pliers B-17025.   16

11. From underneath chassis using creeper 76 replace gearbox drain plug Y-24452 and tighten it up using wrench B-17021.

12. Remove gearbox filler plug and pour in sufficient gear oil to bring it up to level of filler hole. Replace filler plug.

    SPECIAL NOTE.—Gearbox cannot be further progressed until engine is installed in the chassis. It is, therefore, necessary at this point to proceed with installing engine and clutch (Section 1D, Operation 1).

Fig. 16.

Section 3

REAR AXLE

- A. To remove rear axle from chassis.
- B. To dismantle rear axle.
- C. To re-assemble rear axle.
- D. To install rear axle into chassis.

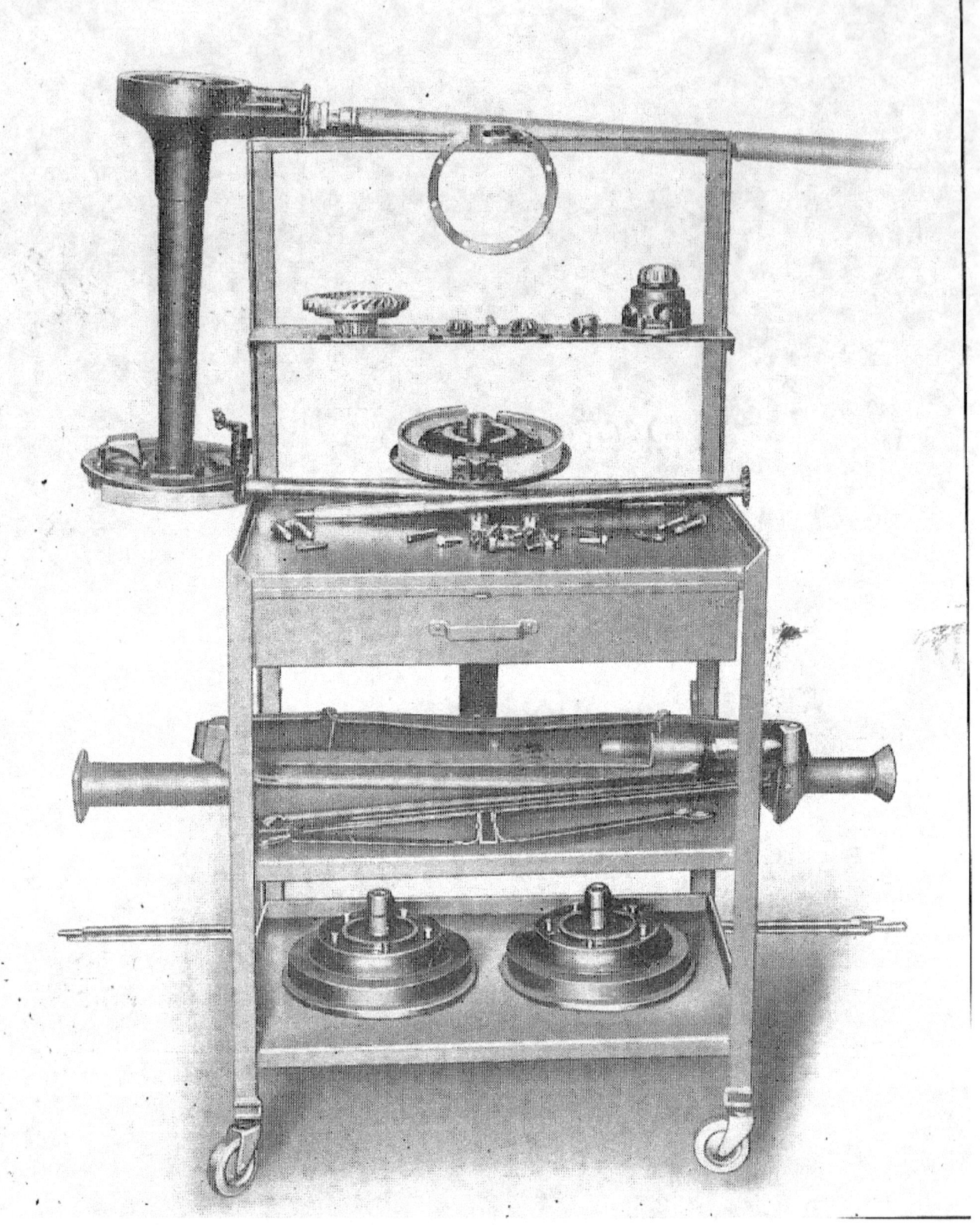

3A

TO REMOVE REAR AXLE FROM CHASSIS

Special Tools and Equipment Required

Tools from Standard Tool Kit

Wrench $\frac{7}{16}''$ and $\frac{1}{2}''$	B-17015
,, $\frac{9}{16}''$ and $\frac{5}{8}''$	B-17016
Screw-driver	B-17020
Pliers	B-17025

Special Tools and Equipment previously used

Hoist	73
Creeper	76
Copper hammer	83
Partition tray	344
Parts carrier	398
Socket	2118
,,	2120
Wrench	2242
Bar handle	2256
"T" wrench	2263
Extension	2297
Cap for gear box cover	—
One drain pan	127

Special Tools and Equipment not previously used

Rear axle stand	50
Two rear axle stand adapters	Y-50-A
Two axle stands	161
Socket	2124

TO REMOVE REAR AXLE FROM CHASSIS

Carry out in sequence the following operations :—

Operation   Illustration

1. Place drain pan 127 under rear axle and remove axle housing drain plug 353053-S using rear end of of adjustable wrench B-17021 which has been ground square in order to fit squared recess formed in plug body.

2. Lift off-side of bonnet and disconnect battery by removing two cover fastening nuts Y-110932 with screw-driver B-17020, slackening off battery negative terminal clamp nut with wrench B-17015 and lifting terminal off battery post. Replace battery cover and nuts loosely.

3. Remove eight screws holding off-side No. 1 sloping floor board YF-940130-B in position with screw-driver B-17020. Slacken lock nut holding accelerator pedal pad YE-11471 in position with wrench B-17015, screw off pedal pad and lift out No. 1 floor board.

4. Remove three screws holding off-side No. 2 horizontal floor board YF-940132-B in position and lift out floor board.

5. Remove six screws holding near-side No. 1 sloping floor board YF-940131-B in position and lift out floor board.

6. Remove three screws holding near-side No. 2 horizontal floor board YF-940133-B in position and lift out floor board.

7. Remove two screws holding foot dimmer switch Y-110378 to floor board plate YF-940016 with screw-driver B-17020, and push switch down. Remove two screws holding front of floor board plate to dash with screw-driver B-17020.
Remove three screws holding rear of floor board plate to cross member with screw-driver B-17020. Slacken off locking ring Y-7228 below gear change lever cap Y-7220

Fig. 17.

Fig. 18.

Operation	Illustration

by tapping with screw-driver B-17020 in slot in ring. Screw off cap Y-7220 and lift out gear change lever. Fit dummy cap on gear change housing in place of Y-7220 to prevent foreign matter entering gearbox. Lift off floor board plate YF-940016.

8. Remove locking wire from universal joint housing cap screw heads using pliers B-17025.

9. From underneath chassis, using creeper 76, remove eight split pins and clevis pins from front end of four rear brake rods Y-2499 and YE-2500-B, rear end of two front brake rods Y-2495, rear end of pedal shaft lever to cross shaft rod Y-2465 and hand brake lever to cross shaft rod YE-2853 using pliers B-17025.   17A

> *NOTE.—The eight clevis pins will be found by frame brake shaft bracket Y-110990 which is located beneath centre frame cross member YR-5025.*

10. From underneath chassis, using creeper 76, remove four split pins, nuts and screws from frame brake shaft bracket Y-110990 to frame centre cross member YR-5025 using wrench 2263, bar 2256, extension 2297 and socket 2118. To release frame brake shaft bracket Y-110990 from frame centre cross member, ease bracket back from flange of cross member which action will free bracket.   17B

11. From underneath chassis, using creeper 76, remove speedometer gear and cap assembly, which is located at front end of torque tube Y-4505-A, using wrench B-17015.   17C

12. From underneath chassis, using creeper 76, remove remaining bolt from universal joint housing cap using wrench 2263, bar 2256, extension 2298 and socket 2120.

13. Remove three upper bolts from universal joint housing cap using wrench 2263, bar 2256 and socket 2120.

13A. Replace drain plug 353053-S in rear axle housing and tighten.

3A

Operation *Illustration*

14. From underneath chassis, using creeper 76, remove nut from bolt which passes through rubber bushed joint of shock absorber body. Wrench B-17016 should be used for this purpose. Links may now be eased off bolt and allowed to hang loose. Other shock absorber should be treated in similar manner.   18A

15. From underneath chassis, using creeper 76, remove four split pins, four nuts and two bars from rear spring clips Y-5705, using pliers B-17025, wrench 2263, bar handle 2256, extension 2297 and socket 2120.   18B

16. Raise chassis at rear end using hoist 73 so as to allow rear axle to be wheeled out.

17. Wheel rear axle out from under chassis and lower chassis down on two axle stands 161 previously placed in position, remove hoist 73.

18. Raise rear axle by means of hoist 73 and place on rear axle stand 50 using adapters 50-A to clamp axle to stand 50.

19. Remove both road wheels using wrench 2242 and socket 2124.

ILLUSTRATIONS AND MEMORANDA

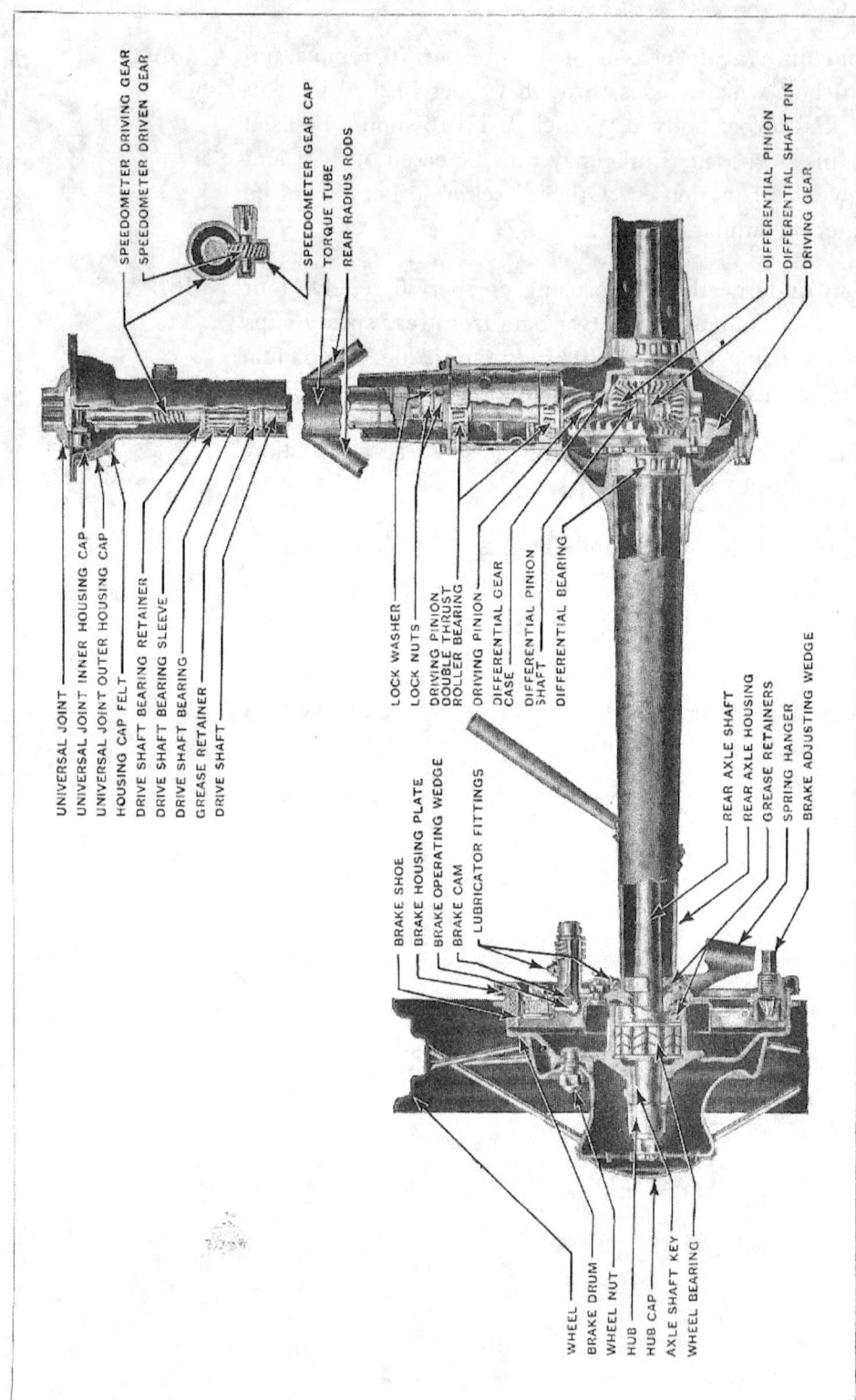

Rear Axle.

3B

TO DISMANTLE REAR AXLE

Special Tools and Equipment Required

Tools from Standard Tool Kit

Wrench $\frac{7}{16}''$ and $\frac{1}{2}''$	B-17015
,, $\frac{9}{16}''$ and $\frac{5}{8}''$	B-17016
,, adjustable	B-17021
Pliers	B-17025

Special Tools and Equipment previously used

Rear axle stand	50
Two rear axle stand adapters	Y-50-A
Copper hammer	83
Partition tray	344
Brass drift	382
Parts carrier	398
Wrench	1616
Socket	2118
,,	2120
Wrench	2242

Special Tools and Equipment not previously used

Rear hub and brake drum puller	Y-115
Rear axle shaft nut wrench	Y-115-N
,, spring expander	Y-321
Differential housing holder	ABV-394
,, ,, ,, adapter plate	Y-394

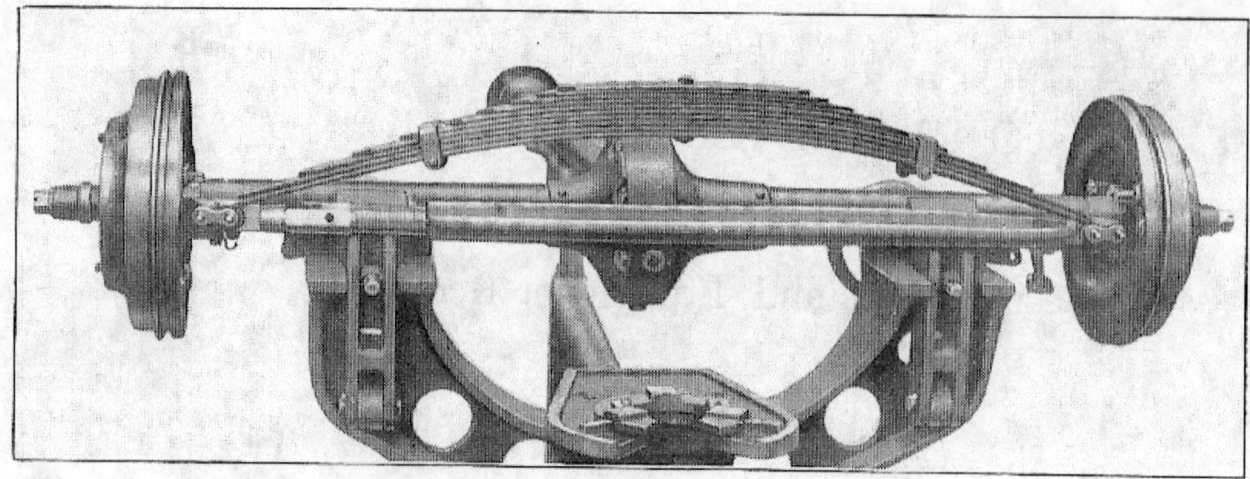

Fig. 19.

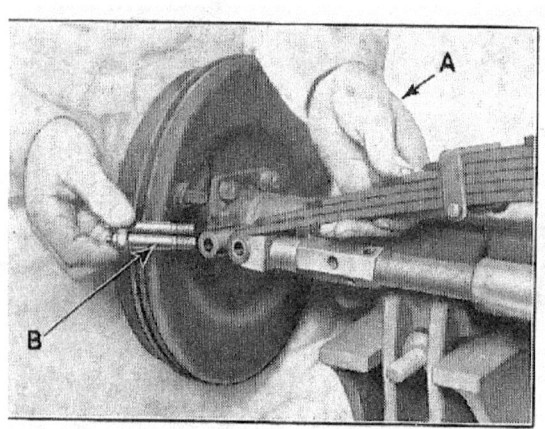

Fig. 20.

Fig. 21.

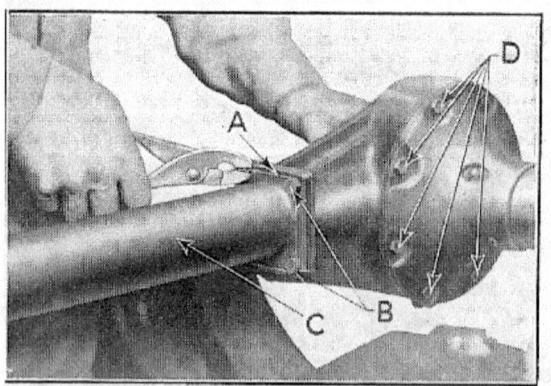

Fig. 22.

3B

TO DISMANTLE REAR AXLE

Carry out in sequence the following operations :—

Operation		Illustration
1.	Expand rear spring Y-5560 using spring expander Y-321.	19

   NOTE.—In carrying out this operation care must be exercised that expander Y-321 is assembled correctly to the rear spring before expanding it. Spring should not be expanded more than necessary to take tension off shackles.

2. Remove split pins, nuts and bars Y-5718 from shackles using pliers B-17025, wrench 2242 and socket 2120.

3. Remove shackles from spring, holding spring to prevent it from falling.   20A

   NOTE.—Shackles should pull out by hand, if operation 2 has been carefully carried out.

4. Remove split pins from outer ends of axle shafts Y-4235 using pliers B-17025.

5. Remove nuts from outer ends of axle shafts Y-4235 using wrench Y-115-N.

6. Remove hub and brake drum assemblies Y-1115 using puller Y-115.   21

7. Remove keys Y-4243 from keyways in axle shafts Y-4235.

8. Remove split pin from bolt securing front ends of rear radius rods Y-4750 and Y-4751 using pliers B-17025.

9. Remove locking wire from four bolt heads at rear end of torque tube Y-4505-A using pliers B-17025.   22

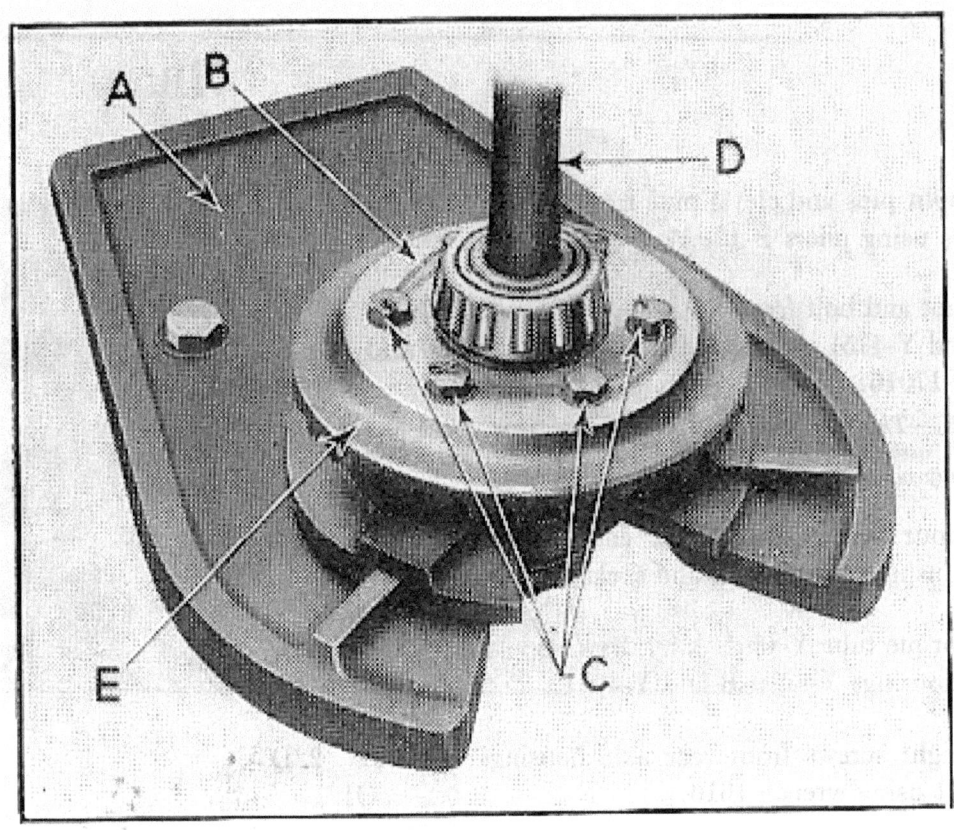

Fig. 25.

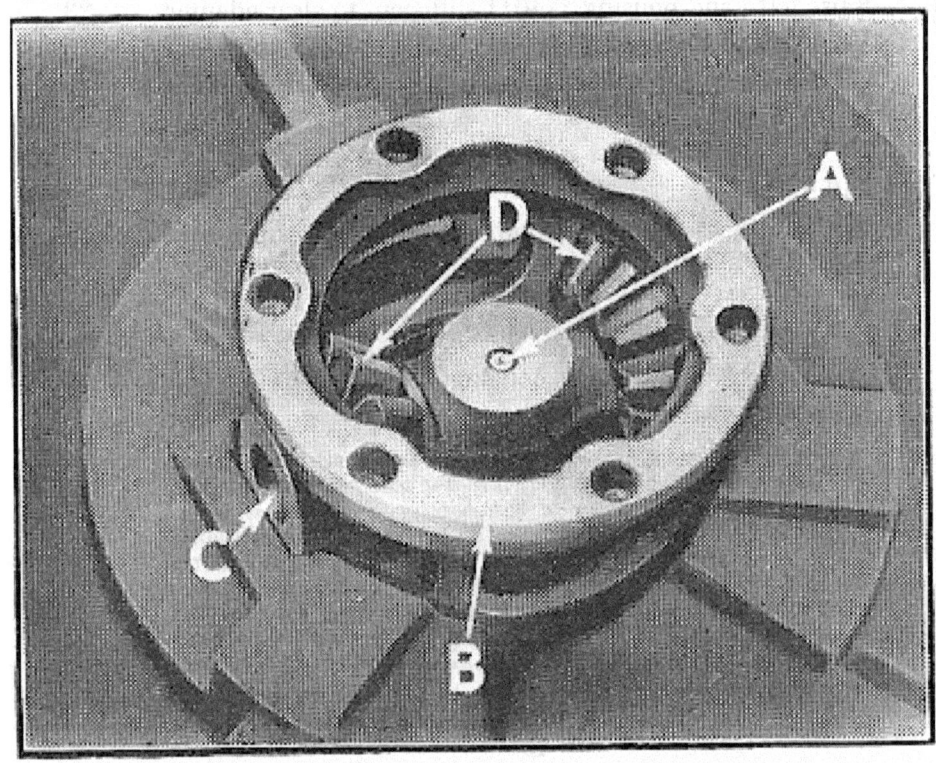

Fig. 26.

3B

Operation		Illustration
19.	Place axle shaft Y-4235 and differential gear case Y-4205-B assembly on adapter plate Y-394 with six differential gear case screw heads uppermost.	25A

> *NOTE.—Differential housing holder A.B.V. 394 will be found bolted to rear axle stand 50 and adapter plate Y-394 should be laid on holder A.B.V. 394.*

20.	Remove locking wire from six differential gear case screw heads using pliers B-17025.	25B
21.	Remove six differential gear case screws using wrench 1616.	25C
22.	Lift off axle shaft Y-4235 and gear YE-4209-D now released by operation 22.	25D
23.	Remove gear YE-4209-D from axle shaft Y-4235 by sliding gear towards tapered end of shaft.	25E
24.	Remove differential gear case Y-4205-B assembly from adapter plate Y-394 by taking hold of remaining axle shaft Y-4235 and inverting assembly. Pin Y-4218 is retained in position by plain central portion of axle shaft bevel gears. It is not secured in any other way. When axle shaft is removed and the assembly inverted this pin should fall out of bushing Y-4212. If it does not do so it is permissible to tap tapered end of axle shaft Y-4235 which will dislodge pin from its position. Copper hammer 83 should be used in this operation.	26A
25.	Replace differential gear case Y-4205-B assembly on adapter plate Y-394 with pinions Y-4215 uppermost.	26B
26.	Remove differential pinion shaft Y-4211 by pushing it out of differential gear case Y-4205-B.	26C

3B

Operation *Illustration*

27. Remove two differential pinions Y-4215 and bushing   26D
    Y-4212 released by operation 27.

28. Remove differential gear case Y-4205-B from axle shaft
    Y-4235 by sliding case towards tapered end of shaft.

TO RE-ASSEMBLE REAR AXLE

Special Tools and Equipment Required

Tools from Standard Tool Kit

Wrench $\frac{7}{16}''$ and $\frac{1}{2}''$	B-17015
,, $\frac{9}{16}''$ and $\frac{5}{8}''$	B-17016
,, adjustable	B-17021
Pliers	B-17025

Special Tools and Equipment previously used

Rear axle stand	50
,, ,, ,, adapters	Y-50-A
Hoist	73
Rear axle shaft nut wrench	Y-115-N
Rear spring expander	Y-321
Partition tray	344
Differential housing holder	ABV-394
,, ,, ,, adapter plate	Y-394
Parts carrier	398
Wrench	1616
Socket	2118
,,	2120
,,	2124
Wrench	2242

Fig. 26.

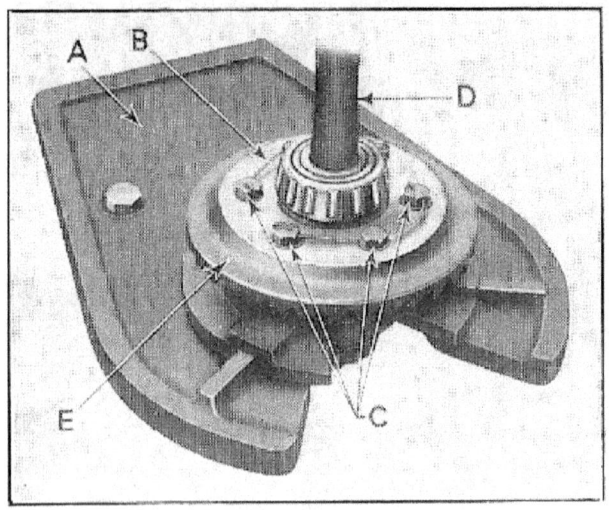

Fig. 25.

Fig. 24.

3c

TO RE-ASSEMBLE REAR AXLE

Carry out in sequence the following operations :—

Operation		Illustration
1.	Place differential gear case Y-4205-B on adapter plate Y-394 with six screw holes uppermost.	26B
2.	Insert axle shaft Y-4235 tapered end first into differential gear case Y-4205-B.	
	NOTE.—The two axle shafts Y-4235 are identical and interchangeable.	
3.	Replace two differential pinions Y-4215 and bushing Y-4212 with differential gear case Y-4205-B.	26D
4.	Replace differential pinion shaft Y-4211, this shaft is a push fit.	26C
	NOTE.—In assembling pinion shaft Y-4211 note that small hole drilled in centre comes in line with corresponding small hole in bushing Y-4212.	
5.	Replace differential pinion shaft pin Y-4218.	26A
6.	Insert axle shaft Y-4235 tapered end first into gear YE-4209-D.	
7.	Offer up axle shaft Y-4235 and gear YE-4209-D assembly on differential gear case Y-4205-B.	
8.	Replace six screws through gear Y-4209-D into differential gear case Y-4205-B and tighten them up evenly using wrench 1616. These screws should be locked by threading a length of wire through each hole in screw heads locking ends of wire together.	25C & B
9.	Remove axle shaft Y-4235 and differential gear case Y-4205-B assembly from adapter plate Y-394.	
10.	Replace rear axle housing Y-4010-B and drive shaft Y-4605-B assembly on stand 50 leaving clamp Y-50-A loose.	24

Fig. 27.

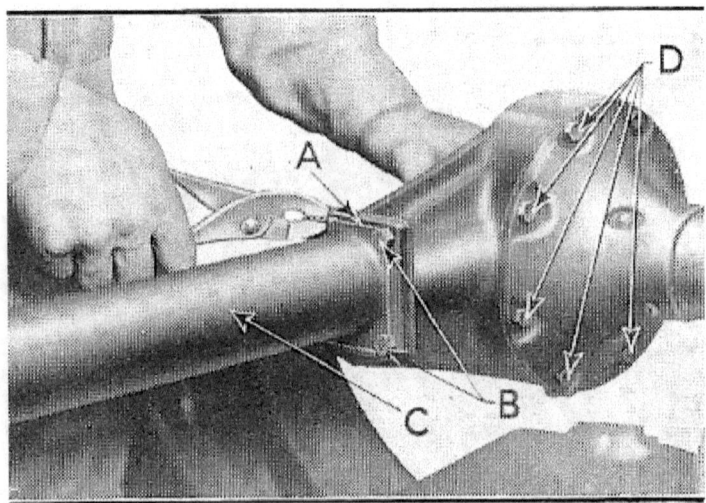

Fig. 22.

Operation		Illustration
11.	Insert axle shaft Y-4235 and differential gear case Y-4205-B assembly into rear axle housing Y-4010-B.	27

NOTE.—This assembly should be inserted into case so that toothed side of gear YE-4209-D is towards housing Y-4010-B. Teeth of gear YE-4209-D will then automatically engage with teeth of pinion.

12.	Secure rear axle housing gasket Y-4035 to face of rear axle housing Y-4010-B by smearing gasket with grease on one side.	
13.	Replace rear axle housing Y-4011.	
14.	Replace eight screws through housing Y-4011 into housing Y-4010-B and tighten them up evenly using wrench 1616.	22D
15.	Secure torque tube to rear axle housing gasket Y-4507 to housing Y-4010-B by smearing gasket with grease on one side.	
16.	Replace torque tube Y-4505-A.	

NOTE.—Torque tube should be replaced with rear radius rod Y-4750 lug underneath.

17.	Replace four screws at torque tube Y-4505-A to rear axle housing Y-4010-B and tighten them up evenly using wrench 2242 and socket 2118. These screws should be locked by threading length of wire through hole in each screw head, locking ends of wire together.	22B & A
18.	Tighten up clamps Y-50-A.	
19.	Offer up four brake rods and support assembly to lug located under torque tube Y-4505-A.	
20.	Align holes of rear radius rods Y-4750 and rear brake rod support Y-110346 with hole in lug on torque tube Y-4505-A, replace bolt and run on nut, tighten by using wrench 2242, socket 2120 and wrench B-17016. This nut should be split pinned in position using pliers B-17025.	

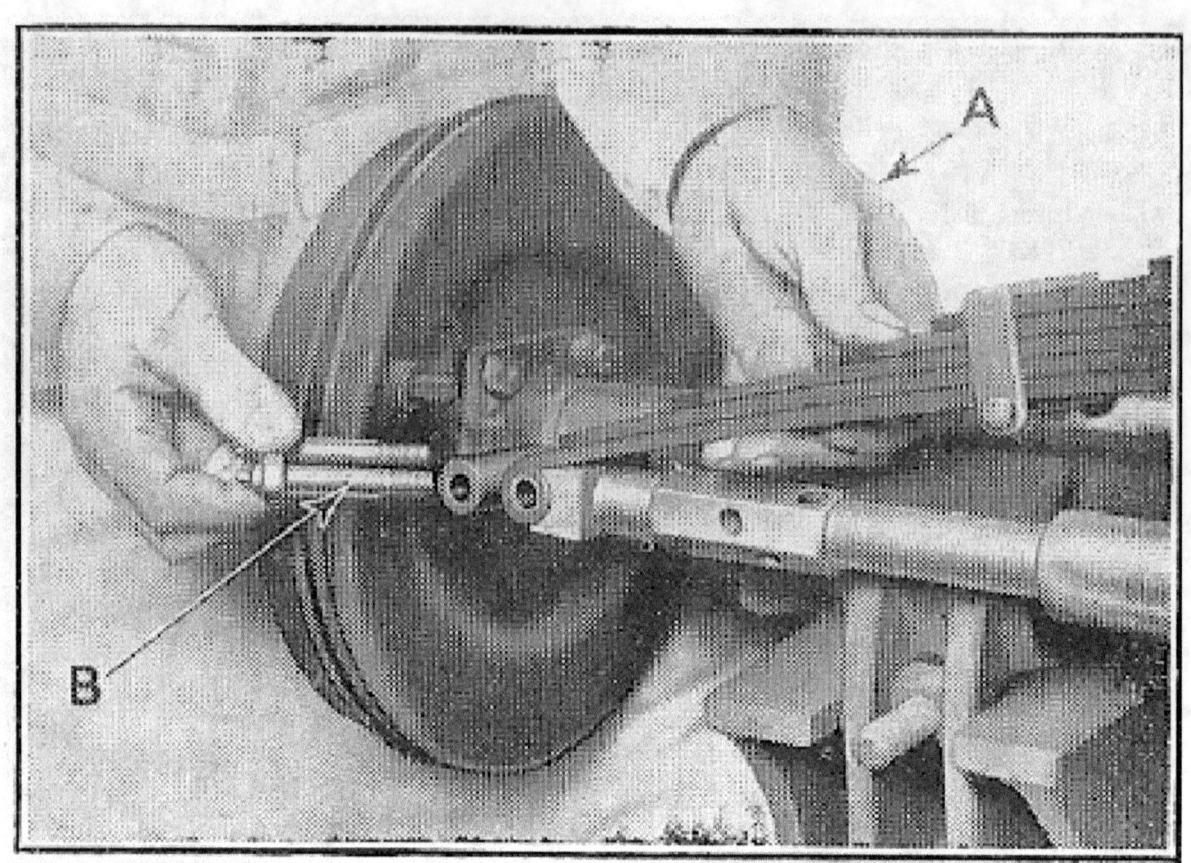

Fig. 20.

Operation	Illustration

21. Replace four clevis pins in rear ends of brake rods and split pin them using pliers B-17025.

22. Replace rear axle shaft keys Y-4243 in their keyways.

23. Replace rear hub and brake drum Y-1115 assemblies on rear axle shafts Y-4235 making quite certain that keyways in drums are in line with keys in shafts.

24. Check carefully to ensure that rear hub gasket B-1183 has not fallen out from outer end of taper on rear hub and brake drum Y-1115 assembly.

25. Replace plain washer 351505-S and rear axle shaft nut 34034-S18 and tighten up nut using wrench Y-115-N. This nut should be split pinned in position using pliers B-17025.

26. Offer up rear spring Y-5560 assembly and insert shackles Y-5715,   20B

27. Replace shackle bars Y-5718 and run on nuts tightening them up using wrench 2242 and socket 2120. These nuts should be split-pinned in position using pliers B-17025.

28. Remove rear spring expander Y-321.

29. Replace rear axle housing drain plug using wrench B-17021.

30. Replace both rear wheels and nuts using wrench 2242 and socket 2124.

31. Raise rear axle by means of hoist 73 from stand 50 and lower axle to floor.

32. Check rear wheel nuts for tightness using wrench 2242 and socket 2124.

TO INSTALL REAR AXLE INTO CHASSIS

Special Tools and Equipment Required

Tools from Standard Tool Kit

Wrench $\frac{7}{16}''$ and $\frac{1}{2}''$	B-17015
,, $\frac{9}{16}''$ and $\frac{5}{8}''$	B-17016
Screw-driver	B-17020
Wrench adjustable	B-17021
Pliers	B-17025

Special Tools and Equipment previously used

Hoist	73
Creeper	76
Partition tray	344
Parts carrier	398
Socket	2118
,,	2120
Bar handle	2256
"T" wrench	2263
Extension	2297

Fig. 18.

Fig. 17

TO INSTALL REAR AXLE INTO CHASSIS

Carry out in sequence the following operations :—

Operation | Illustration

1. With hoist 73 raise chassis from rear and remove two axle stands 161.

2. Wheel rear axle under chassis.

3. Raise front end of torque tube Y-4505-A and offer into position allowing it to rest on frame centre cross member YR-5025.

4. From underneath chassis, using creeper 76, locate the rear spring Y-5560 assembly in rear frame cross member Y-5030 lowering chassis to spring by means of hoist 73.

5. From underneath chassis, using creeper 76, replace spring clip bars Y-5712 and run on nuts, tightening them up evenly using wrench 2263, bar handle 2256, extension 2297 and socket 2120. These nuts should be split-pinned in position using pliers B-17025. 18B

6. From underneath chassis, using creeper 76, replace two rear shock absorber links to shock absorber bodies, run on the nuts and tighten them using wrench B-17016. 18A

7. From underneath chassis, using creeper 76, replace speedometer gear and cap assembly Y-17270. Secure by tightening screws using wrench B-17015. 17C

8. From underneath chassis, using creeper 76, replace frame brake shaft bracket Y-110990 on frame centre cross member and bolt it up, using wrench 2263, bar handle 2256, extension 2297 and socket 2118. Nuts should be split-pinned in position using pliers B-17025.

ILLUSTRATIONS AND MEMORANDA

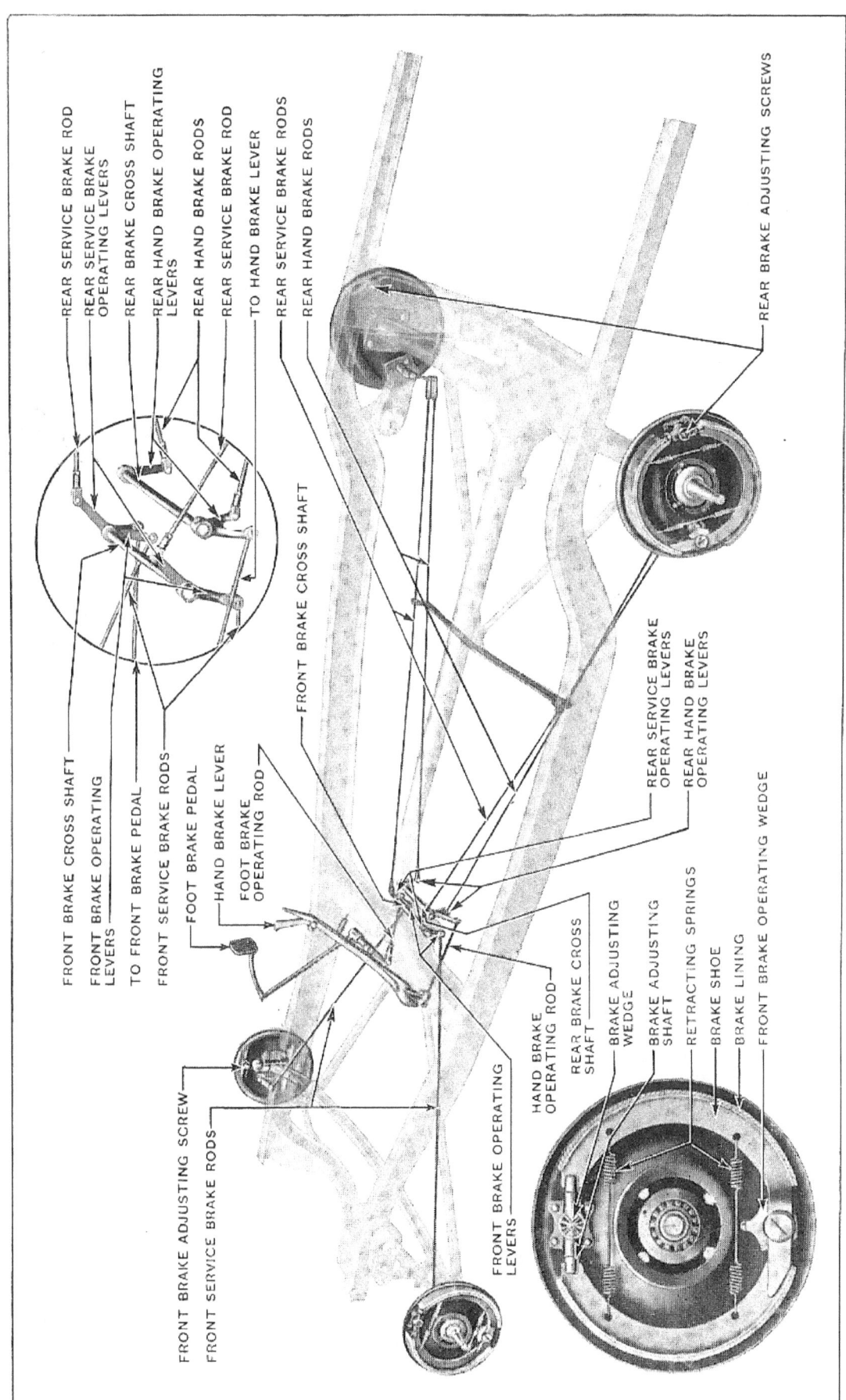

Brake Layout.

Operation   *Illustration*

9. Replace eight clevis pins in frame brake shaft bracket arms, i.e., four in front ends of rear brake rods, two in rear end of front brake rods, one in rear of pedal shaft lever to cross shaft rod and one in rear of hand brake lever to cross shaft rod. These clevis pins should be split-pinned in position using pliers B-17025.   17A

10. Remove filler plug 353053-S in rear axle housing with wrench B-17021 and refill axle with oil to level of filler plug. Replace filler plug and tighten using wrench B-17021.

Section 4

FRONT AXLE

A. To remove front axle from chassis.

B. To dismantle front axle.

C. To re-assemble front axle.

D. To install front axle into chassis.

TO REMOVE FRONT AXLE FROM CHASSIS

Special Tools and Equipment Required

Tools from Standard Tool Kit

Wrench $\frac{9}{16}''$ and $\frac{5}{8}''$	B-17016
Pliers	B-17025

Special Tools and Equipment previously used

Rear axle stand	50
Rear axle stand adapters	50-A
Hoist	73
Creeper	76
Copper hammer	83
Axle stands (two)	161
Socket	2118
,,	2120
,,	2124
Wrench	2242
Bar handle	2256
" T " wrench	2263
Extension	2297

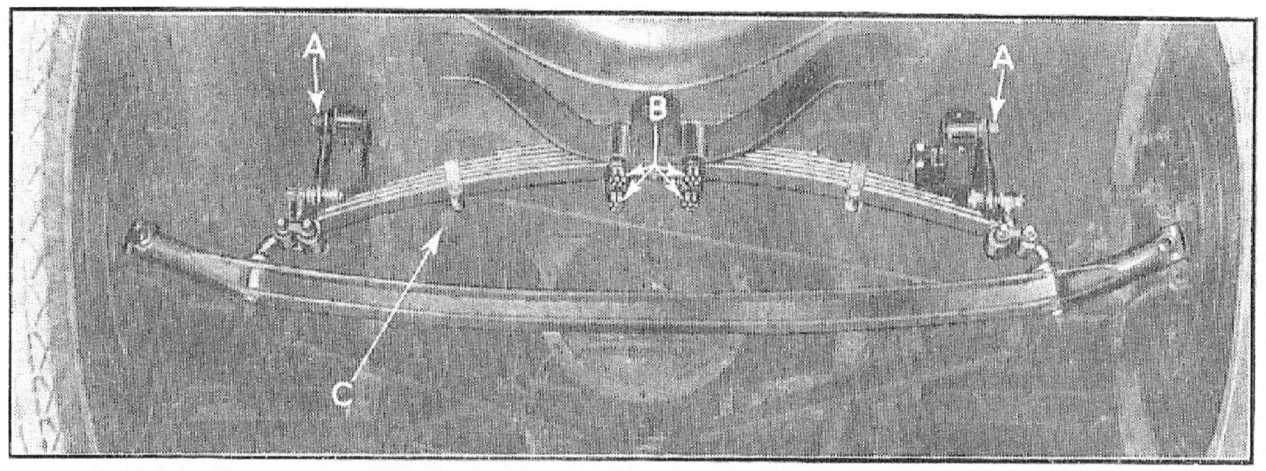

Fig. 28.

Fig. 17.

TO REMOVE FRONT AXLE FROM CHASSIS

Carry out in sequence the following operations :—

Operation		Illustration
1.	Loosen twelve wheel hub bolt nuts Y-1012, using wrench 2242 and socket 2124.	
2.	Remove nut from bolt which passes through rubber bushed joint of the off-side shock absorber body using wrench B-17016. Link may now be eased off bolt and allowed to hang loose. Near-side shock absorber should be treated in similar manner.	28A
3.	From underneath chassis, using creeper 76, remove four split pins, four nuts and two bars from front spring clips Y-5455 using pliers B-17025, wrench 2263, bar handle 2256, extension 2297 and socket 2120.	28B
4.	From underneath chassis, using creeper 76, remove four split pins, nuts and bolts from frame brake shaft bracket Y-110990 to frame centre cross member YR-5025, using wrench 2263, bar handle 2256, extension 2297 and socket 2118.	17B
5.	From underneath chassis, using creeper 76, remove stop-light tension spring with pliers B-17025, from rear end of off-side front brake rod Y-2495. This spring may be unhooked from one end only and allowed to hang loose.	
6.	From underneath chassis, using creeper 76, remove two split pins and clevis pins from rear ends of off-side and near-side front brake rods Y-2495.	
7.	From underneath chassis, using creeper 76, remove frame brake shaft bracket Y-110990 from frame centre cross member YR-5025 by easing bracket back from flange of cross member which action will free bracket.	

Operation		Illustration
8.	From underneath chassis, using creeper 76, remove split pin and nut from rear end of drag link assembly YE-3304-A stud using pliers B-17025 and wrench B-17016.	28C

> NOTE.—*Drag link assembly stud will have to be given a smart tap at its threaded end in order to release it from steering gear arm YE-3590-A. Copper hammer 83 should be used for this operation.*

9. Raise chassis from front end using hoist 73 so as to allow front axle to be wheeled out.

10. Wheel front axle out from under chassis and lower chassis down on two axle stands 161 previously placed in position. Remove hoist 73 from chassis.

11. Raise front axle by means of hoist 73 and place on rear axle stand 50 using adapters 50-A to clamp axle to stand.

12. Remove both road wheels using wrench 2242 and socket 2124.

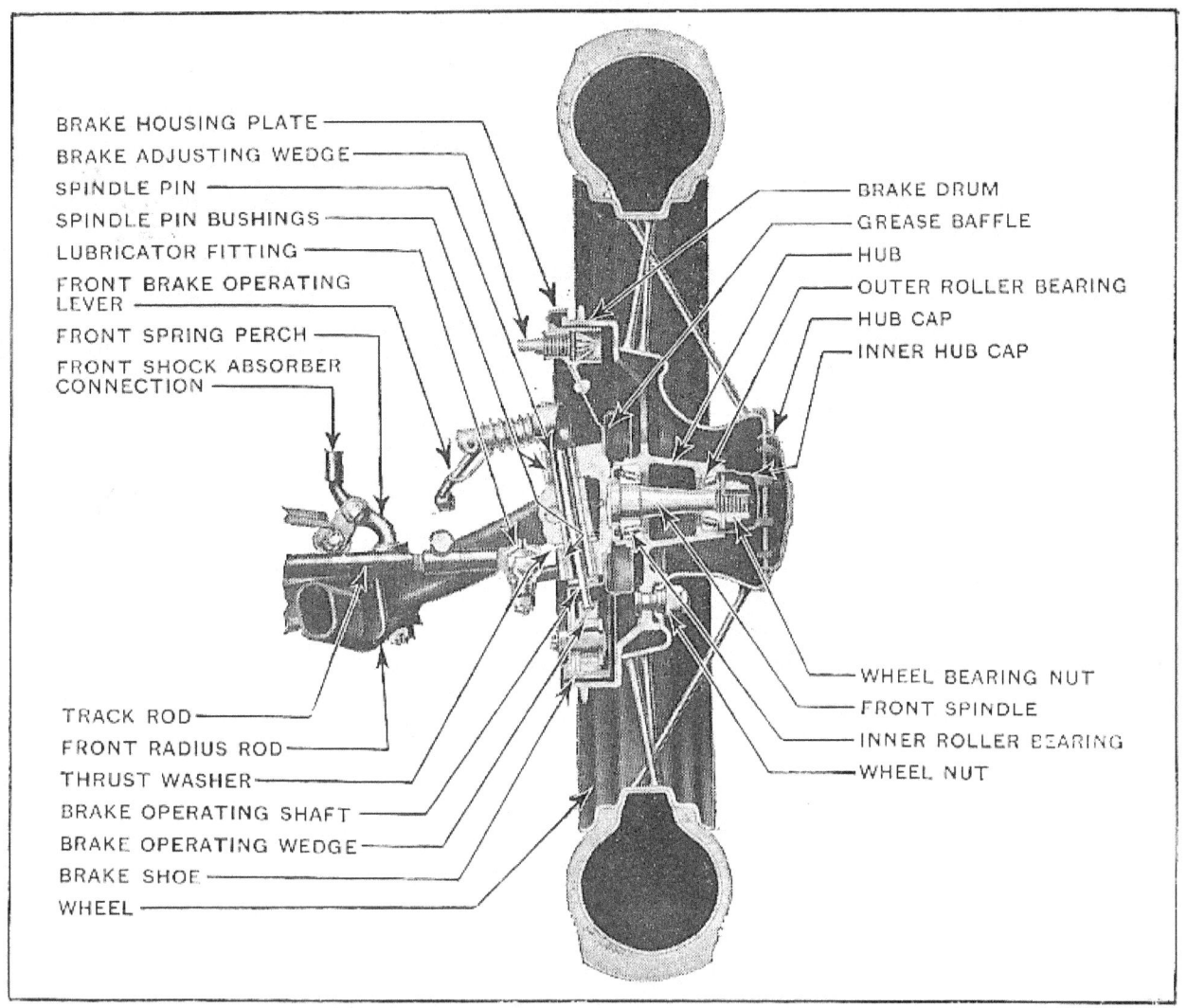

Front Hub and Brake Section

TO DISMANTLE FRONT AXLE

Special Tools and Equipment Required

Tools from Standard Tool Kit

Wrench $\frac{7}{16}''$ and $\frac{1}{2}''$	B-17015
,, $\frac{9}{16}''$ and $\frac{5}{8}''$	B-17016
Pliers	B-17025

Special Tools and Equipment previously used

Copper hammer	83
Wrench	Y-115-N

4B

TO DISMANTLE FRONT AXLE

Carry out in sequence the following operations :—

> NOTE.— *Each of the following operations applies to the near-side and off-side components of front axle assembly.*

Operation   Illustration

1. Remove split pins and clevis pins from front ends of brake rods Y-2495, using pliers B-17025 and release rods from supports YE-2502.

2. Remove front hub grease caps Y-1139 using wrench Y-115-N.

3. Remove split pins from spindle nuts 34034-S using pliers B-17025.

4. Remove spindle nuts 30034-S using wrench Y-115-N.

5. Draw off front hub and brake drum assemblies Y-1105, noting that hub grease retainer washers Y-1195 and outer front wheel bearing cone assemblies Y-1216 are removed at same time.

6. Remove inner front wheel bearing cone assemblies Y-1201 from spindle assemblies Y-3105-6.

7. Remove split pin and nut from spindle connecting rod tube assemblies YE-3281 using pliers B-17025 and wrench B-17016.

    > NOTE.—*The spindle connecting rod end studs will have to be given a smart tap at their threaded end in order to release them from the spindle assemblies Y-3105-6. Copper hammer 83 should be used for this operation.*

8. Remove lower spindle assembly lubricator fittings using wrench B-17015.

Operation *Illustration*

9. Remove four split pins nuts and bolts holding front brake housing plate assemblies Y-2011 to spindle assemblies Y-3105-6 using pliers B-17025, wrench B-17016 and wrench 1616.

   NOTE.—In removing brake housing plates it will be necessary to hold front brake operating pins Y-2075 to facilitate withdrawal of brake housing plates.

10. Remove spindle bolt lock pin nuts Y-3124 using wrench B-17016.

11. Remove spindle bolt lock pins Y-3122, tapping them out with copper hammer 83.

12. Draw out spindle bolts Y-110983 thus releasing bushed spindle assemblies Y-3105-6.

    NOTE.—In removing bushed spindle assemblies Y-3105-6 be careful that spindle thrust washers Y-3123 which are located between bottom spindle bearings and front axle Y-3010 are not mislaid.

TO RE-ASSEMBLE FRONT AXLE

Special Tools and Equipment Required

Tools from Standard Tool Kit

Wrench $\frac{7}{16}''$ and $\frac{1}{2}''$	B-17015
,, $\frac{9}{16}''$ and $\frac{5}{8}''$	B-17016
Pliers	B-17025

Special Tools and Equipment previously used

Hoist	73
Copper hammer	83
Wrench	Y-115-N
Socket	2124
Wrench	2242

TO RE-ASSEMBLE FRONT AXLE

Carry out in sequence the following operations:—

> NOTE.—Each of the following operations applies to the near-side and off-side components of the front axle assembly.

Operation   *Illustration*

1. Offer up spindle assemblies Y-3105-6 to front axle Y-3010 and insert spindle thrust washers Y-3123 between bottom spindle bearings and front axle Y-3010.

2. Replace spindle bolts Y-110983.

3. Replace spindle bolt lock pins Y-3122 and nuts Y-3124 tightening nut up with wrench B-17016.

4. Replace front brake operating pins Y-2075 and felt washers B-3121 and hold in position.

5. Offer up front brake housing plate assemblies Y-2011 to spindle assemblies Y-3105-6 at same time guiding front brake operating pins Y-2075 through holes provided in housing plates, making sure that tops of operating pins Y-2075 are engaged with brake shaft cam sockets in spindle bolt heads and lower ends located securely in socket of brake operating wedges Y-2050 between brake shoes Y-2019.

6. Offer up front brake grease baffle assemblies Y-2060 to front brake housing plate assemblies Y-2011 and insert four bolts holding these two assemblies to spindle assemblies Y-3105-6, run on the four nuts and tighten them up using wrenches B-17016 and 1616. These nuts should be split pinned in position using pliers B-17025.

7. Replace lower spindle assembly lubrication fittings using wrench B-17015.

4C

Operation *Illustration*

8. Replace spindle connecting rod tube assemblies YE-3281 and run on nuts, tightening them up with wrench B-17016. Nuts should be split pinned in position using pliers B-17025.

9. Replace inner front wheel bearing cone assemblies Y-1201 on spindle assemblies Y-3105-6.

10. Replace front hub and brake drum assemblies Y-1105 on spindle assemblies Y-3105-6.

11. Replace outer front wheel bearing cone assemblies Y-1216 on spindle assemblies Y-3105-6.

12. Replace hub grease retainer washers Y-1195 and spindle nuts 30034-S using wrench Y-115-N.

    NOTE.—Spindle nuts should be tightened up so that hub and brake drum assemblies Y-1105 just rotate and then backed off one half turn.

13. Replace front hub grease caps Y-1139 tightening them up with wrench Y-115-N.

14. Replace brake rods Y-2495 in supports YE-2502 and insert clevis pins in front ends at front brake levers Y-2084 split pinning them in position using pliers B-17025.

15. Replace both front wheels and nuts using wrench 2242 and socket 2124.

16. Raise front axle by means of hoist 73 from stand 50 and lower axle to floor.

17. Check front wheel nuts for tightness using wrench 2242 and socket 2124.

TO INSTALL FRONT AXLE INTO CHASSIS

Special Tools and Equipment are Required

Tools from Standard Tool Kit

Wrench $\tfrac{9}{16}''$ and $\tfrac{5}{8}''$	B-17016
Pliers	B-17025

Special Tools and Equipment previously used

Hoist	73
Creeper	76
Axle stands (two)	161
Socket	2118
,,	2120
Bar handle	2256
"T" wrench	2263
Extension	2297

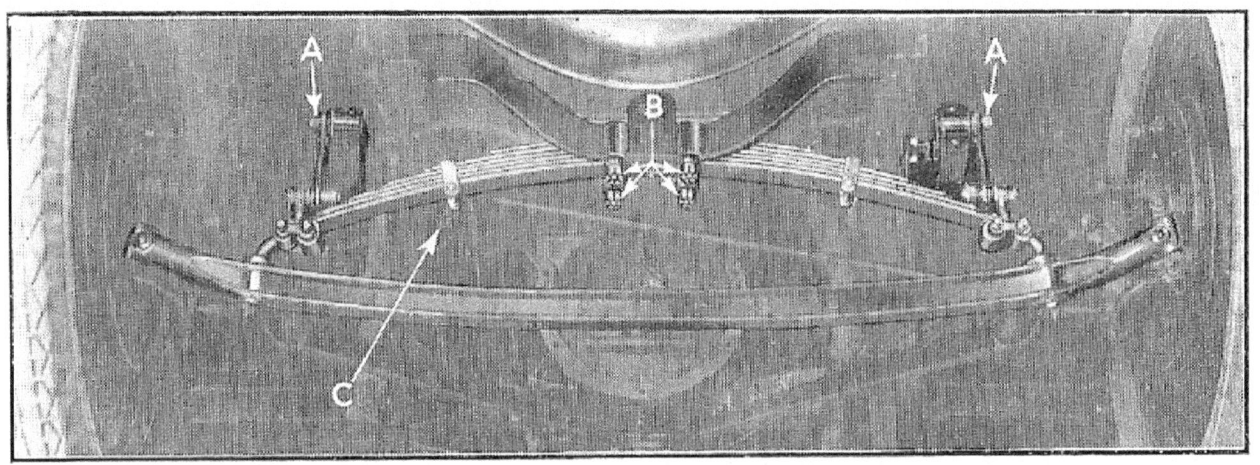

Fig. 28.

TO INSTALL FRONT AXLE INTO CHASSIS

Carry out in sequence the following operations :—

Operation		Illustration
1.	With hoist 73 raise chassis from front and remove two axle stands 161.	
2.	Wheel front axle under chassis.	
3.	From underneath chassis, using creeper 76, locate front spring YE-5310 assembly in frame cross member Y-5020, lowering chassis to spring by means of hoist 73. When in position remove hoist 73.	
4.	From underneath chassis, using creeper 76, replace two spring clip bars Y-5458 and run on four nuts tightening them up diagonally using wrench 2263, bar handle 2256, extension 2297 and socket 2120. These nuts should be split pinned in position using pliers B-17025.	28B
5.	From underneath chassis, using creeper 76, replace the frame brake shaft bracket Y-110990 on frame centre cross member YR-5025 and bolt it in position using wrench 2263, bar handle 2256, extension 2297 and socket 2118. These nuts should be split pinned in position using pliers B-17025.	
6.	From underneath chassis, using creeper 76, replace two clevis pins in rear ends of two front brake rods Y-2495 and split pin them, replace stop-light tension spring on rear end of off-side brake rod using pliers B-17025.	
7.	From underneath chassis, using creeper 76, replace drag link assembly YE-3304-A stud in tapered hole of steering gear arm YE-3599-A, run on the nut and tighten it up using wrench B-17016. This nut should be split pinned in position using pliers B-17025.	28C
8.	Replace two front shock-absorber links to shock-absorber bodies, run on nuts and tighten them up using wrench B-17016.	28A

Section 5

STEERING GEAR

A. To remove steering gear from chassis.

B. To dismantle steering gear.

C. To re-assemble steering gear.

D. To install steering gear in chassis.

TO REMOVE STEERING GEAR FROM CHASSIS

Special Tools and Equipment Required

Tools from Standard Tool Kit

Wrench $\frac{7}{16}''$ and $\frac{1}{2}''$	B-17015
,, $\frac{9}{16}''$ and $\frac{5}{8}''$	B-17016
Screw-driver	B-17020
Pliers	B-17025

Special Tools and Equipment previously used

Rear axle stand	50
Creeper	76

Special Tools and Equipment not previously used

Steering gear arm puller	Y-345-P
Steering wheel puller	Y-373
Steering wheel insert pliers	Y-373-B
Steering wheel nut wrench	Y-854-A
Rear axle stand adapters	Special

Fig. 29.

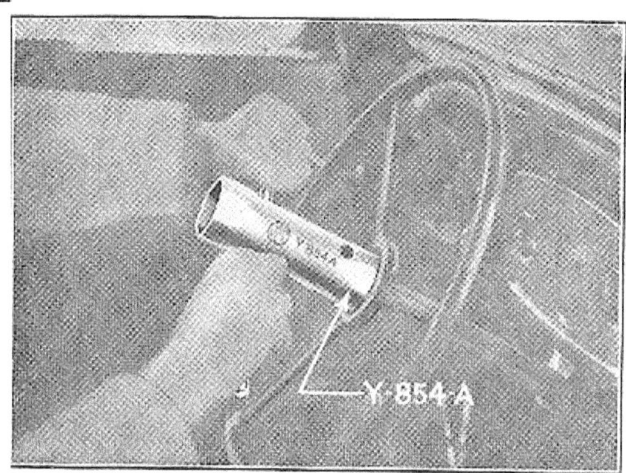

Fig. 30.

TO REMOVE STEERING GEAR FROM CHASSIS

Carry out in sequence the following operations :—

Operation		Illustration
1.	Lift off-side of bonnet and disconnect battery by removing two cover fastening nuts Y-110932 with screw-driver B-17020, slacking off battery negative terminal clamp nut with wrench B-17015 and lifting terminal off battery post. Replace battery cover and nuts loosely.	
2.	Remove steering wheel insert Y-110823 from centre of steering wheel YE-3600-C using special pliers Y-373-B.	29
3.	Remove horn switch and nut assembly YE-3616-B using wrench Y-854-A.	

NOTE.—The horn switch wire connection is of "push-in" type and when nut has been unscrewed this connection should be pulled out. This also applies to connection at lower end of steering assembly and which may now be treated in a similar manner.

4.	Remove steering wheel YE-3600-C using puller Y-373.	30

NOTE.—Do not remove steering wheel key 74178-S.

5.	Remove two steering column support clip screws using screw-driver B-17020.	
6.	Remove screw holding front of off-side engine pan Y-110291-B to front cross member Y-110072 with screw-driver B-17020.	
7.	From underneath car, using creeper 76, remove two bolts and nuts holding off-side engine pan Y-110291-B to side member with wrench B-17015 and remove pan.	
8.	From underneath car, using creeper 76, remove split pin and nut from steering shaft Y-110845 using pliers B-17025 and wrench B-17016.	

5A

Operation   *Illustration*

9. From underneath car, using creeper 76, remove steering gear arm YE-3590-A from lower steering shaft Y-110845 using puller Y-345-P.

10. From underneath car, using creeper 76, remove steering gear arm key and felt washer from lower steering shaft Y-110845.

11. Remove off-side shock absorber from chassis frame only. This is secured in position by two bolts passing through flange of shock absorber and chassis frame. Use wrench B-17016 for this operation.

12. Remove split pins from two bolts securing steering gear housing to upper flange of chassis frame using pliers B-17025.

13. Remove three bolts securing steering gear housing to chassis frame using wrenches B-17015 and B-17016.

14. Remove four screws securing two halves of draught excluder in position to scuttle-dash using screw-driver B-17020.

15. Remove steering gear assembly YE-3503-A from chassis and place on rear axle stand 50 using special adapters.

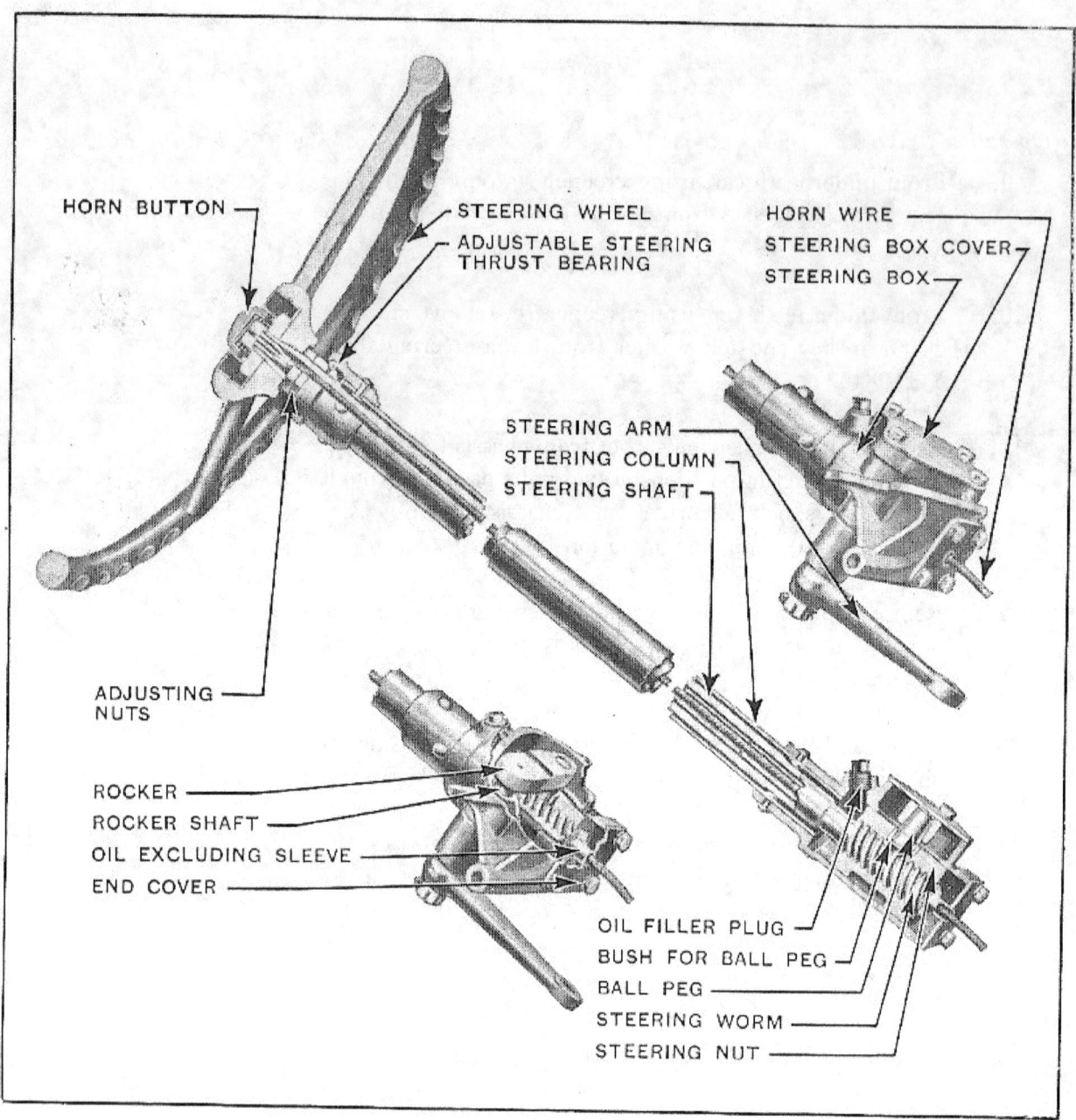

Steering Gear.

TO DISMANTLE STEERING GEAR

Special Tools and Equipment Required

Tools from Standard Tool Kit

Wrench $\frac{7}{16}''$ and $\frac{1}{2}''$ B-17015
Screw-driver B-17020

Special Tools and Equipment not previously used

Steering lock-nut wrenches (two) CY-345-L

Special Tools and Equipment previously used

Drain Pan.. 127

DISMANTLING THE STEERING GEAR

Carry out in sequence the following operations:—

Operation   Illustration

1. Remove horn switch wire assembly YE-14308 which passes through centre of upper steering gear shaft Y-110827. This wire may be drawn out by hand from steering wheel end of shaft Y-110827.

2. Remove four screws from steering gear housing cover YE-3580 using wrench B-17015. This will allow oil to run out which must be caught in Drain Pan 127.

3. Remove steering gear housing cover YE-3580 and gasket YE-3581.

4. Remove lower steering shaft and ball peg assembly Y-110845 from steering gear housing.
   NOTE.—This assembly will push out by hand.

5. Remove four screws from steering gear housing end plate and sleeve assembly Y-110844 using wrench B-17015.

6. Remove steering gear housing end plate and sleeve assembly Y-110844 and gasket Y-110837.

7. Replace steering wheel YE-3600-C on upper steering gear shaft.
   NOTE.—There is no need to replace horn switch and nut assembly YE-3616-B.

8. Turn steering wheel YE-3600-C in an anti-clockwise direction and at same time hold " main nut " at bottom end of upper steering gear shaft Y-110827 to prevent rotation. This action will run " main nut " off upper steering gear shaft allowing it to be withdrawn from steering gear housing.

9. With two wrenches CY-345-L unlock nuts located under steering wheel YE-3600-C.

5B

Operation *Illustration*

10. Remove steering wheel YE-3600 and key 74178-S using screw-driver B-17020 to ease out key.

11. Run off steering gear ball race lock-nut Y-110839 and upper steering shaft ball race adjusting nut YE-3517.

12. Remove sixteen steel balls from steering gear ball race.

13. Withdraw upper steering gear shaft Y-110827-B from steering column tube and housing assembly Y-110843.

TO RE-ASSEMBLE STEERING GEAR

Special Tools and Equipment Required

Tools from Standard Tool Kit

Wrench $\frac{7}{16}''$ and $\frac{1}{2}''$ B-17015

Special Tools and Equipment previously used

Rear axle stand 50
Rear axle stand adapters .. .. Special
Steering lock-nut wrenches (two) CY-345-L

TO RE-ASSEMBLE STEERING GEAR

Carry out in sequence the following operations :—

Operation   Illustration

1. Replace upper steering gear shaft Y-110827-B in steering column tube and housing assembly Y-110843.

2. Replace sixteen steel balls in steering gear ball race.

    NOTE.—The steel balls may be coated with grease to keep them in position when assembling.

3. Replace upper steering shaft ball race adjusting nut YE-3517, and run on lock-nut Y-110839.

4. Replace steering wheel key 74178-S and offer steering wheel YE-3600-C into position.

5. Replace " main nut " on lower end of upper steering gear shaft Y-110844 by turning steering wheel YE-3600-C in clock-wise direction, holding " main nut " to prevent rotation.

    NOTE.—" Main nut " must be assembled with side which has largest amount of cutaway towards off-side frame member YE-5015-B.

6. Replace steering gear housing end plate and sleeve assembly Y-110844 and gasket Y-110837.

7. Replace four screws in steering gear housing end plate and tighten them up using wrench B-17015.

8. Replace lower steering shaft and ball peg assembly Y-110845 in steering gear housing, engaging ball peg of steering shaft in recess provided for it in " main nut."

9. Turn steering wheel to bring " main nut " to bottom of its travel and pour into steering gear housing sufficient gear oil to fill it right up.

10. Replace steering gear housing cover YE-3580 and gasket YE-3581.

11. Replace four screws in steering gear housing cover YE-3580 and tighten them up using wrench B-17015.

12. Replace horn switch wire assembly YE-14308 inserting it from lower end of steering column tube and housing assembly Y-110843.

13. Remove filler plug Y-E-3538 with wrench, fill with oil to correct level and replace plug.

TO INSTALL STEERING GEAR IN CHASSIS

Special Tools and Equipment Required

Tools from Standard Tool Kit

Wrench $\frac{7}{16}''$ and $\frac{1}{2}''$	B-17015
,, $\frac{9}{16}''$ and $\frac{5}{8}''$	B-17016
Screw-driver	B-17020
Pliers	B-17025

Special Tools and Equipment previously used

Creeper	76
Steering lock-nut wrenches (two)	CY-345-L
Wrench	Y-853
Steering wheel nut wrench	Y-854-A

5D

TO INSTALL STEERING GEAR IN CHASSIS

Carry out in sequence the following operations :—

Operation   Illustration

1. Offer steering assembly YE-3503-A in position on chassis frame.

2. Replace three steering gear housing to chassis frame bolts and tighten them up using wrenches B-17015 and B-17016. Nuts should be split pinned in position using pliers B-17025.

3. Replace steering column support clip and two screws using screw-driver B-17020.

4. Replace off-side shock absorber and tighten up two bolts using wrench B-17016.

5. Replace draught excluder and four screws securing it to the scuttle-dash using screw-driver B-17020.

6. Replace steering wheel key 74178-S in keyway of upper steering gear shaft Y-110827-B.

7. Replace steering wheel YE-3600-C noting that key 74178-S is positioned correctly.

8. Replace horn switch and nut assembly YE-3616-B and tighten it up using wrench Y-854-A.

    NOTE.—Before screwing nut home on upper steering gear shaft Y-110827 horn switch wire connection should be snapped into position.

9. Replace steering wheel insert Y-110823.

10. Snap lower end of horn switch wire assembly YE-14308 into its connection.

11. From underneath car, using creeper 76, replace steering gear arm key 74178-S in keyway of lower steering shaft Y-110845.

171

Operation *Illustration*

12. From underneath car, using creeper 76, replace steering gear arm YE-3590-A noting that key 74178-S is positioned correctly.

13. From underneath car, using creeper 76, replace and tighten up steering gear arm nut using wrench B-17016. This nut should be split pinned in position using pliers B-17025.

14. From underneath car, using creeper 76, offer up off-side engine pan Y-110291-B. From above replace screw to hold off-side engine pan to front cross member Y-110072 using screwdriver B-17020.

15. From underneath car, using creeper 76, replace two bolts, spring washers, and nuts, to hold off-side engine pan to frame side member using wrenches B-17015 and Y-853. For ease of operation assemble both bolts loosely, starting with rear bolt.

16. Adjust steering ball race nut allowing steering wheel YE-3600-C $\frac{3}{4}''$ free movement measured on rim of the steering wheel. Lock nuts together using wrenches CY-345-L.

17. Replace battery negative terminal lead by removing two cover fastening nuts Y-110932 with screw-driver B-17020 and lift cover. Clamp lead to negative terminal post using wrench B-17015. Coat battery terminal liberally with vaseline to prevent corrosion. Replace battery cover and tighten up fastening nuts.

ILLUSTRATIONS AND MEMORANDA

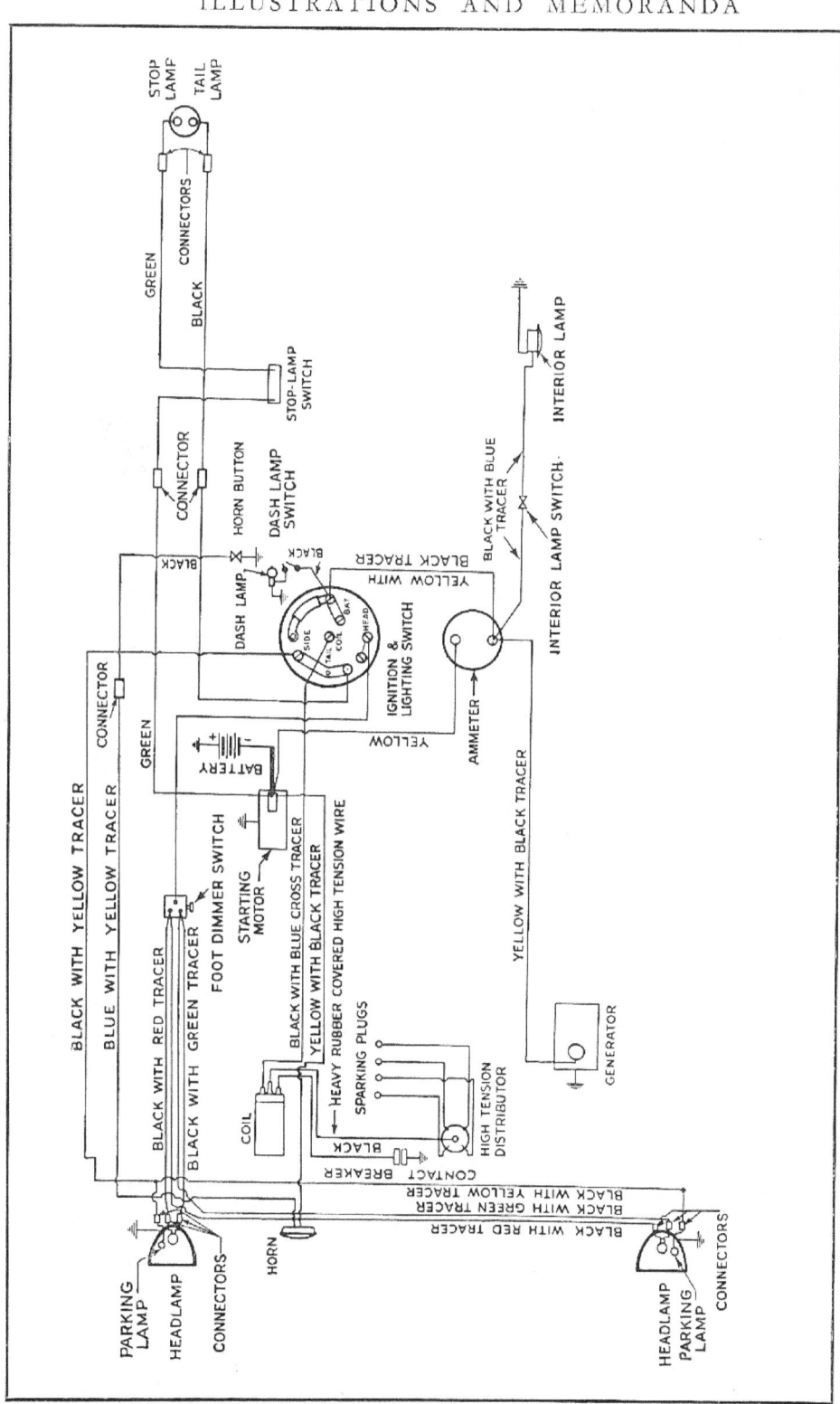

Wiring Diagram.

Section 6

TO DISMANTLE AND RE-ASSEMBLE DISTRIBUTOR, GENERATOR, STARTER MOTOR, CARBURETTOR, AND FUEL PUMP

TO DISMANTLE AND RE-ASSEMBLE DISTRIBUTOR, GENERATOR, STARTER MOTOR, CARBURETTOR, AND FUEL PUMP

Special Tools and Equipment Required

Tools from Standard Tool Kit

Wrench $\frac{7}{16}"$ and $\frac{1}{2}"$	B-17015
,, $\frac{9}{16}"$ and $\frac{5}{8}"$	B-17016
Pliers	B-17025
Screw-driver	B-17020

Special Tools and Equipment previously used

Wrench	Y-810
Wrench	Y-853
Copper Hammer	83

Special Tools and Equipment not previously used

Pin Punch	37
Wrench	5324
Special Wrench for Starter Pinion	Y-858-S

ILLUSTRATIONS AND MEMORANDA

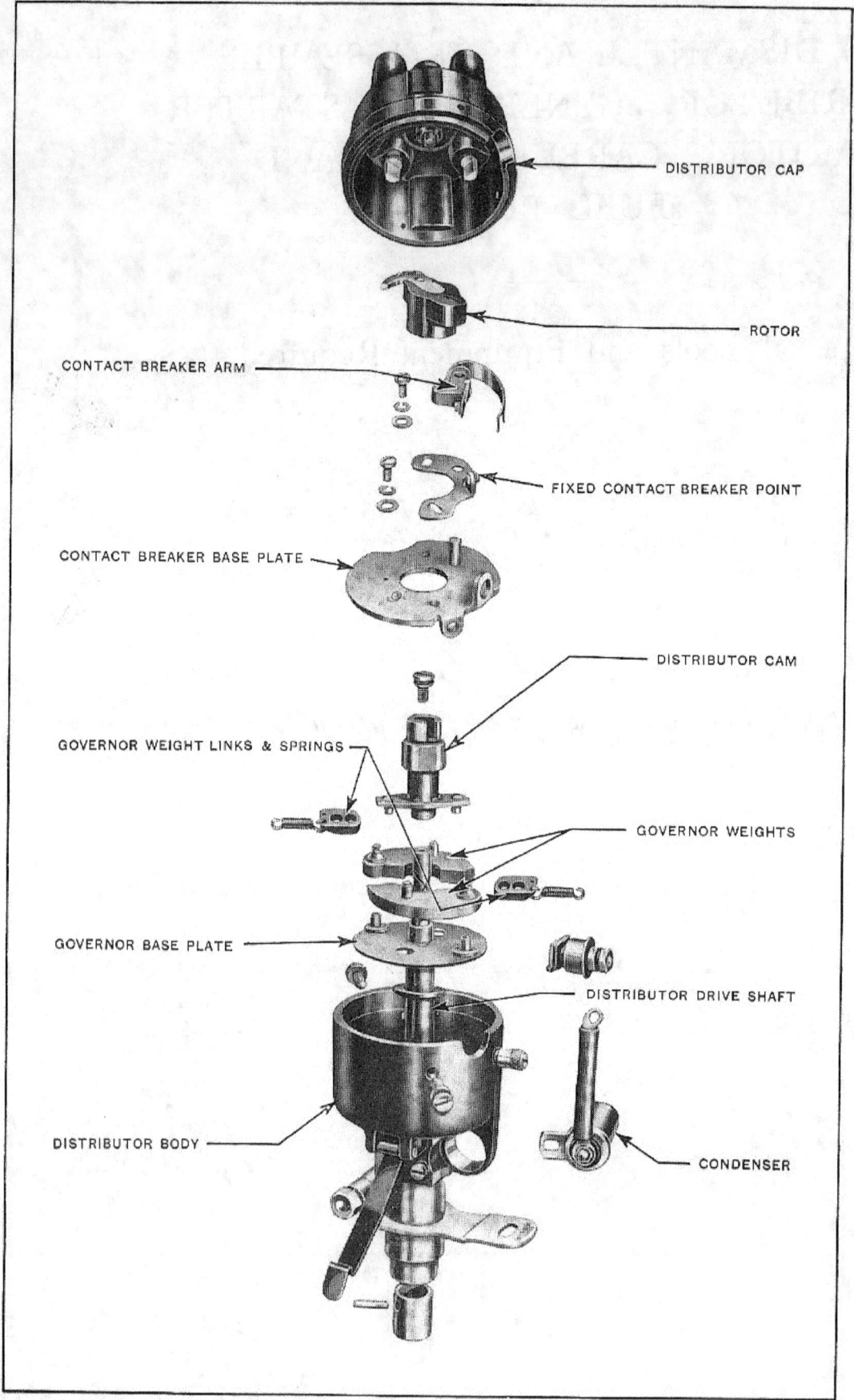

TO DISMANTLE DISTRIBUTOR

Carry out in sequence the following operations:—

Operation *Illustration*

1. Release distributor cap holding down clips and remove cap YE-12116-B from distributor body YE-12130-B.

2. Lift rotor YE-12200-B off centre shaft.

3. Detach condenser YE-12300-B lead from terminal on side of distributor body using wrench Y-810. Replace one nut and spring washer on condenser terminal. Remove screw, clamping condenser bracket to distributor body, and remove condenser and lead from distributor.

4. Loosen nut on condenser terminal using spanner Y-810. Slip movable contact breaker arm YE-12162 from pivot and condenser lead terminal and remove insulating washer from contact breaker arm pivot.

5. Unscrew two base plate securing screws and lift out base plate YE-12150-B with fixed contact breaker arm YE-12160 and condenser terminal attached.

6. Punch out pin 72857-S holding drive shaft sleeve B-12185 to drive shaft, using copper hammer 83 and pin punch 37. This will permit governor and shaft assembly to be removed from distributor body.

    N.B.—Note position of slot in cam relative to tongue in lower end of drive shaft. This is important when reassembling distributor as it affects position of rotor contact point when retiming.

7. Holding shaft firmly, remove cam retaining screw YE-12211-B and brass washer from shaft and lift cam and governor control arm assembly YE-12179 off shaft.

8. Remove governor weights from their pivots on weight carrier plate. Remove links Y-110797 and brass washers from their pivots on weight carriers plate.

TO RE-ASSEMBLE DISTRIBUTOR

Carry out in sequence the following operations :—

Operation   *Illustration*

1. Wash all parts in petrol and dry thoroughly.

2. Smear a thin film of oil over governor weight pivots on carrier plate. Place brass washer on pivot on each governor weight. Reassemble springs to anchorage on governor weights and replace links Y-110797 on weights so that bevelled edge of link is to the bottom, and hole nearest bevel is located on pivot on weight. Note that each spring is located in hole in link Y-110797 which is nearest to drive shaft. Replace governor weights on pivots on carrier plate.

   NOTE.—Two springs Y-110788 and YE-12191 are not identical, one being of heavier gauge than the other, so as to produce the desired ignition advance characteristics.

3. Place distributor cam and control arm YE-12179 on to centre shaft so that two studs on control arm engage in vacant holes in links Y-110797 making sure that slot in cam is in same relative position with drive shaft tongue as before. Holding drive shaft firmly, replace cam, retaining screw YE-12211-B and brass washer.

4. Lightly oil distributor drive shaft and insert drive shaft and governor assembly into distributor body. Place brass washer and drive shaft sleeve B-12195 on drive shaft and secure in place by pin 72857-S using copper hammer 83.

5. Place base plate YE-12150-B in distributor body and secure by two screws and spring washers.

6. Lightly oil contact breaker arm pivot. Replace insulating washer on pivot. Replace movable contact breaker arm YE-12162 on pivot so that two points are together and end of arm spring fits over condenser terminal lead bolt.

Operation *Illustration*

7. Insert condenser Y-12300-B into lug on distributor body. Insert screw and spring washer in condenser bracket and secure to distributor body. Remove nut and spring washer from condenser lead terminal. Replace condenser lead on terminal and replace two nuts and spring washers, using wrench Y-810.

8. With fibre block on high point of cam set contact breaker gap at from .010" to .012" by loosening two holding down screws that secure fixed contact breaker arm YE-12160 to base plate and moving arm until correct gap is obtained.

9. Lightly smear vaseline on cam. Replace rotor YE-12200-B on centre shaft, engaging tongue in slot on cam. Replace cap YE-12116-B and secure by holding down clips.

NOTE.—To retime distributor, proceed as in Section 1.D, Operations 9 and 26.

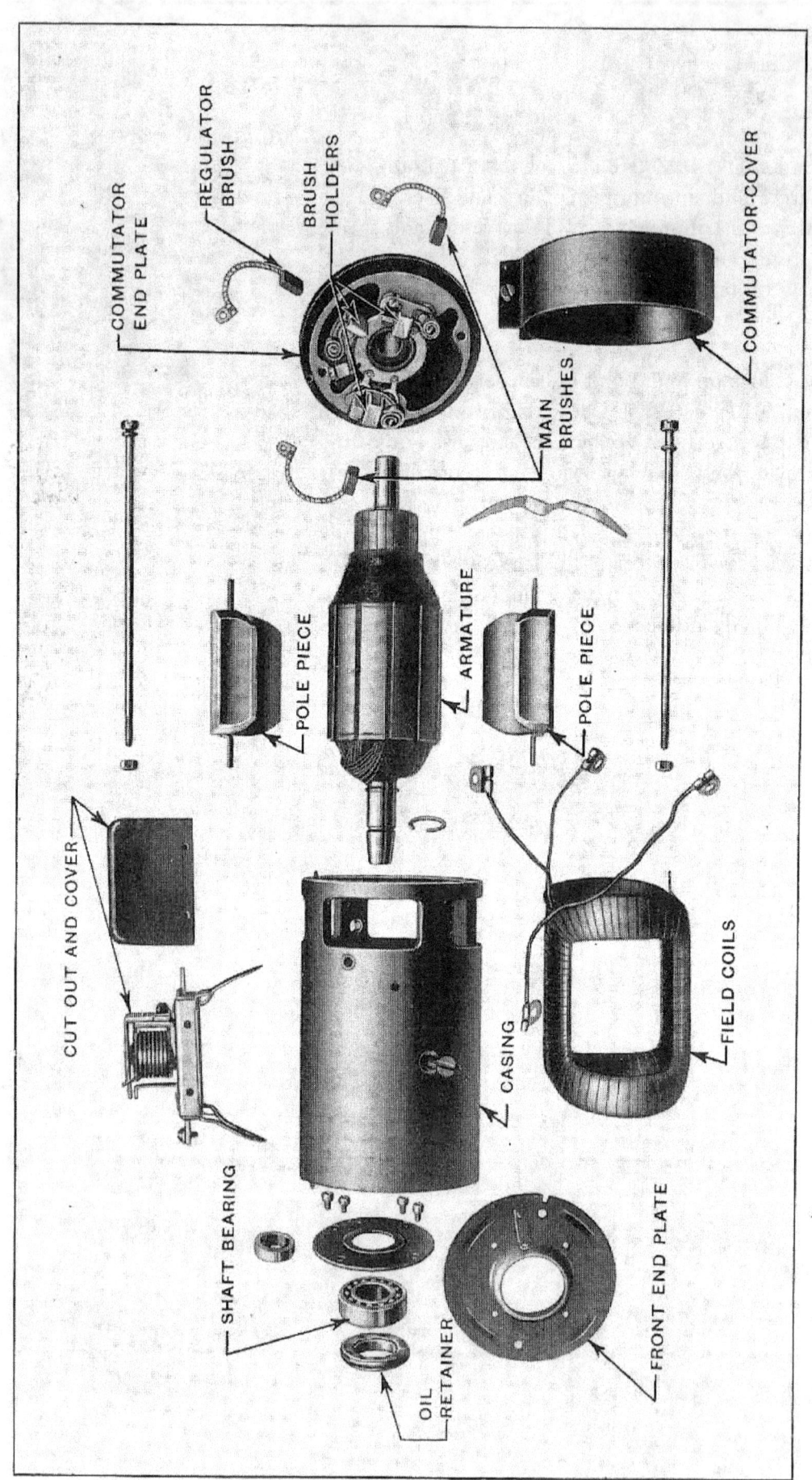

TO DISMANTLE GENERATOR

Carry out in sequence the following operations :—

Operation *Illustration*

1. Remove generator from engine as in Section 1A, Operations 27 and 35.

2. Detach fan Y-8605 and fan pulley YE-8610-B from armature shaft using wrench Y-810.

3. Detach generator commutator band cover YE-10142 from generator body.

4. Disconnect three leads from generator brushes YE-10069 /70.

5. Unscrew two end plate securing nuts, using wrench Y-810, and remove two securing bolts YE-10166 from forward end of generator.

6. Remove two end plates YE-10139 and YE-10050 withdrawing armature YE-10005-B with front end plate. Slide brushes YE-10069/70 out of brush holders on plate YE-10072.

7. Disconnect lead from generator to cutout and remove cutout YE-10505 by unscrewing two holding down screws. (Cutout cover may be pulled off.)

TO RE-ASSEMBLE GENERATOR

Carry out in sequence the following operations :—

Operation *Illustration*

1. Replace armature YE-10005-B and front end plate YE-10139 in generator body, locating end plate slot on dowel on generator body.

2. Replace rear end plate YE-10050 locating hole in end plate on dowel in generator body.

3. Insert two generator main fixing bolts YE-10166 from rear to front and secure by two spring washers and nuts, using wrench Y-810.

4. Replace two main brushes YE-10069 and third brush YE-10070 in brush holders, noting that high point of brush is in direction of rotation.
    NOTE.—To facilitate the operation of replacing brushes, brush holder spring clips should be held back by a piece of bent wire while brush is being inserted.

5. Attach red field lead and brush lead to terminal on bottom main brush holder and secure by fixing screw. Attach two other brush leads and two field leads coloured yellow to terminals on adjacent brush leads.

6. Replace cutout YE-10505 on generator and secure by two holding down screws. Attach red field lead to cutout terminal.

7. Replace generator cover band YE-10142 and secure.

8. Replace generator on engine as in Section 1.D Operation 17.

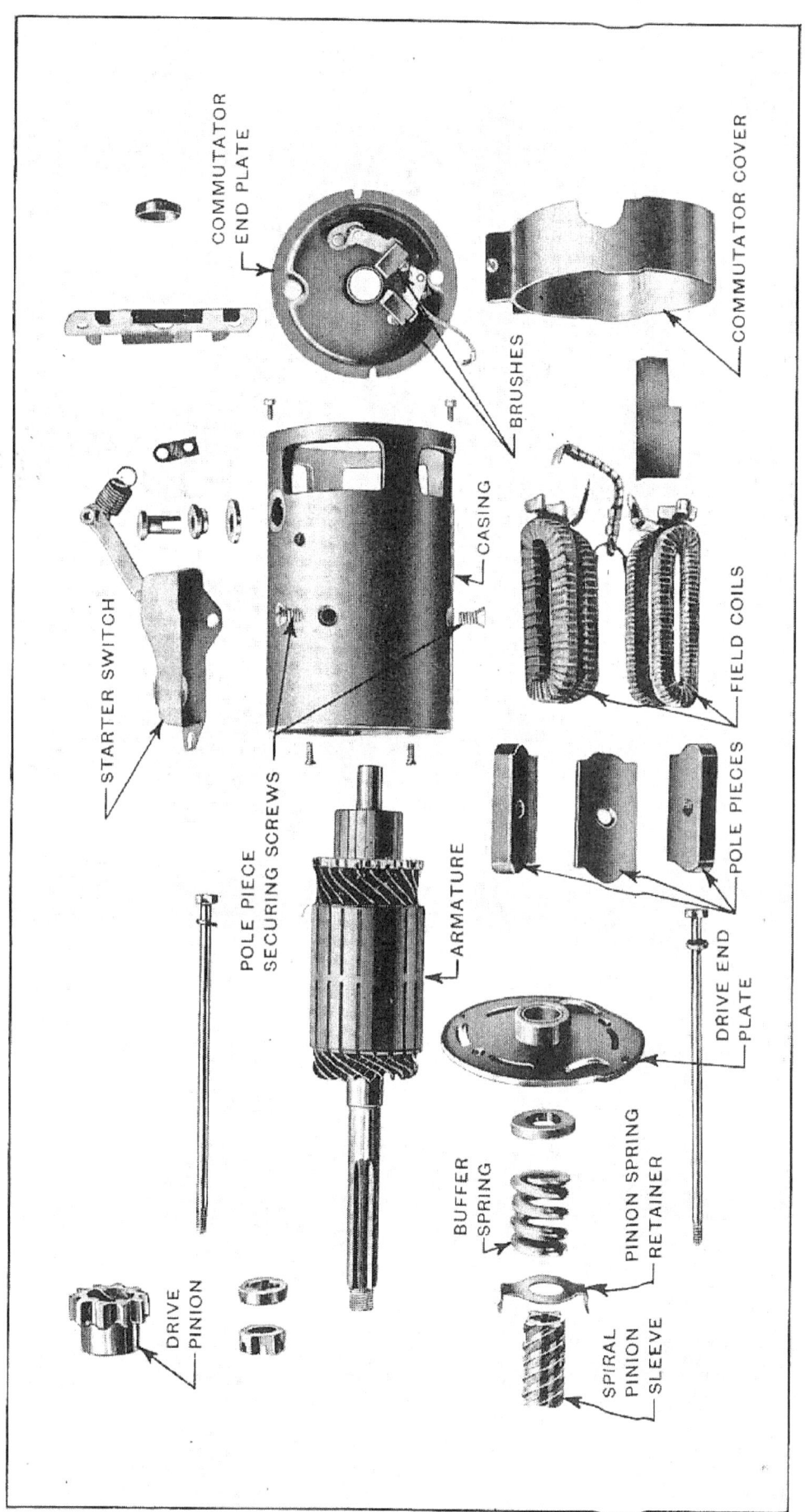

TO DISMANTLE STARTER MOTOR

Carry out in sequence the following operations:—

Operation *Illustration*

1. Remove negative battery lead from battery post.

2. Remove starting switch control cable YE-11475-A at starting switch assembly by unscrewing clamp set screw with wrench 7810.

3. Remove starter motor assembly YE-11000-B from engine as in Section 1A, Operation 31, by unscrewing two long securing bolts YE-11079, with wrench Y-810.

4. Detach starter switch assembly BE-11450 by removing three holding down screws and washers, using screwdriver B-17020.

5. Remove split pin from shaft nut Y-11381, using pliers B-17025 and unscrew nut using wrench 5324 at the same time holding drive pinion YE-11363 with wrench Y-858-S to prevent shaft from turning.

6. Withdraw splined collar Y-110973, pinion YE-11363, sleeve YE-11355, spring clip YE-11359, main spring YE-11375 and sleeve (starter buffer) YE-11362.

7. Remove commutator cover band YE-11126, using screwdriver B-17020, slacken brush screw and remove lead from field coil to brush.

8. Remove rear end plate Y-110965 by unscrewing two counter-sunk screws with screw driver B-17020.

9. Unscrew two end plate, retaining screws, using screwdriver B-17020 and remove end plate Y-110981, at the same time drawing armature from the starter yoke. Lift brushes and remove end plate from armature.

Operation *Illustration*

10. Remove brushes YE-11055 from holders YE-11061 by removing securing screws.

11. Remove brush holder YE-11061 by lifting one end of the brush holder spring from its stud anchorage.

TO RE-ASSEMBLE STARTER MOTOR

Carry out in sequence the following operations:—

Operation *Illustration*

1. Replace brush holder YE-11061 and fix spring in position on its anchorage stud.

2. Replace brushes YE-11055 on holder YE-11061 and secure with screws and spring washers.

3. Replace end plate Y-110981 and brush assembly on to armature shaft, (commutator end) lifting brushes to enable them to slide over commutator.

4. Insert armature Y-11005 in starter yoke and secure end plate Y-110981 in position with two screws using screw-driver B-17020. Connect white field lead to brush. Connect bare copper earth lead to brush.

5. Refit rear end plate Y-110965 and secure by two counter sunk screws, using screw-driver B-17020.

6. Replace sleeve (starter buffer) YE-11362, main spring YE-11375, Spring clip YE-11359, spiral sleeve YE-11355, pinion YE-11363, and splined collar Y-110973. Screw on nut YE-11381, using wrench 5324, at the same time holding pinion with wrench Y-858-S to prevent armature shaft turning. Fit new split pin.

7. Refit commutator cover band and secure by nut and bolt, using screw driver-B-17020.

8. Refit starter switch securing with three screws and washer, using screw driver-B-17020.

9. Replace starter motor assembly on engine securing with two long bolts YE-11079 and bridge piece Y-110997.

10. Attach battery lead to switch terminal.

11. Replace battery negative lead to battery negative post.

TO DISMANTLE CARBURETTOR
Carry out in sequence the following operations :—

Operation   Illustration

1. Disconnect carburettor to accelerator rod Y-9747. This rod has a spring loaded cap and can be pulled off without difficulty.

2. Disconnect carburettor starting control connector from control wire by slackening off screw on connector Y-110850.

3. Remove two nuts and washers, holding carburettor to manifold YE-9425, using wrench Y-853, and lift carburrettor from manifold, preserving gasket Y-9447.

4. Remove screw Y-110861, securing air cowl Y-110860 and lift air cowl from carburettor. Remove air regulating screw Y-110868 and spring Y-110872-B.

5. Remove two screws Y-110865, holding carburettor bowl to barrel. Lift off filter Y-110578. Bowl will drop away.

6. Remove float YE-9555 from bowl.

7. Remove main jet YE-9534-A, compensating jet YE-9575-A, slow running jet YE-9545, and starting jet YE-9594-A from carburettor bowl by unscrewing.

 NOTE.—The emulsion block Y-110881-B should not be disturbed.

8. Remove needle valve and seat Y-110863, using wrench B-17015.

9. Remove starting device control valve assembly C-110554, using wrench B-17015.

10. Remove automatic air valve assembly Y-110855 by unscrewing.

 NOTE.—The throttle lever and plate should not be removed.

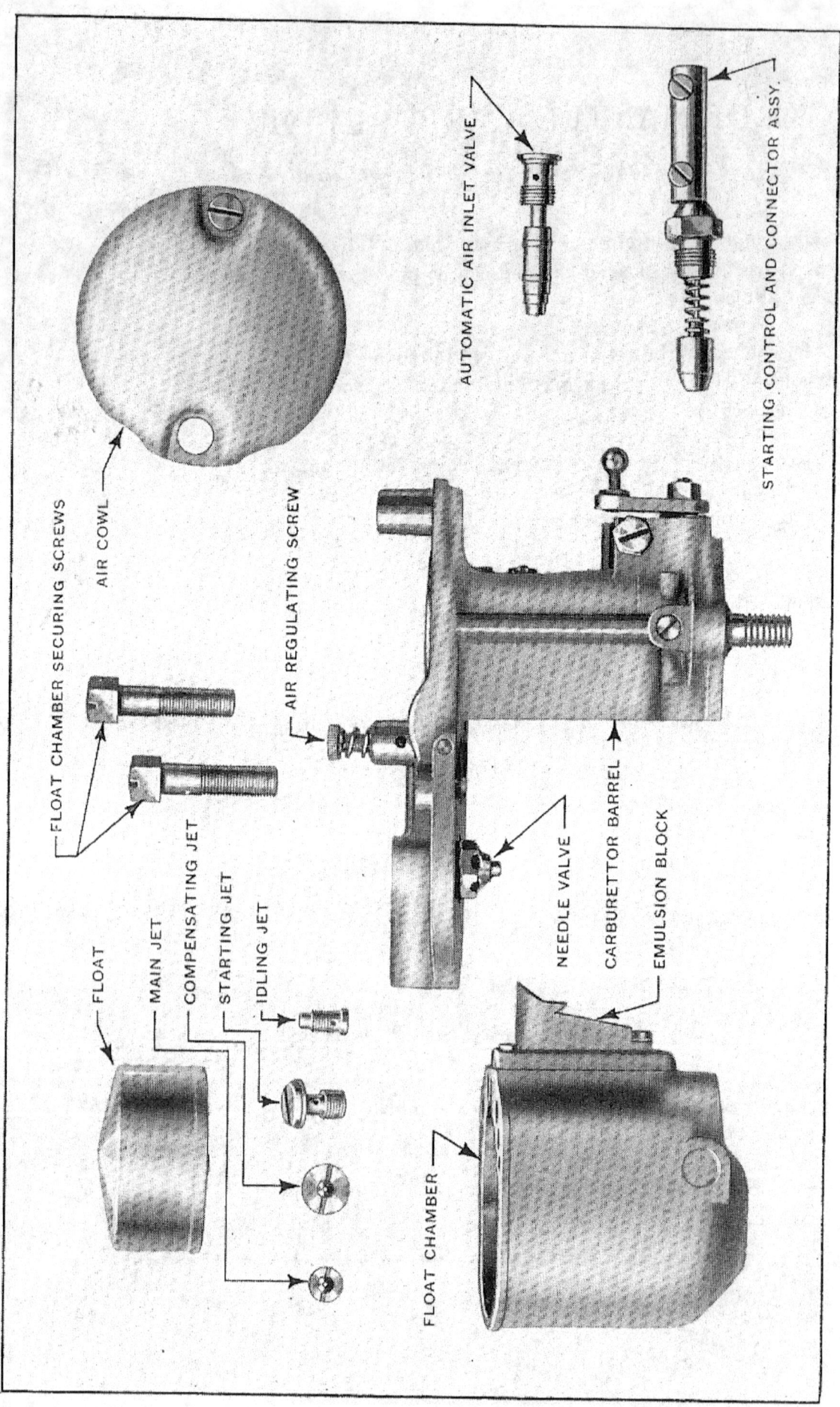

TO RE-ASSEMBLE CARBURETTOR

Carry out in sequence the following operations:—

Operation *Illustration*

1. Insert automatic air valve assembly Y-110855 in its socket and screw home.

2. Insert starting device control valve assembly in its socket and screw home, using wrench B-17015.

3. Replace needle valve and seat Y-110863 into its socket in carburettor barrel head and tighten down carefully, using wrench B-17015.

4. Replace starting jet YE-9594-A, slow running jet YE-9545, compensating jet YE-9575-A and main jet YE-9534-A, in sockets in carburettor bowl.

5. Replace float YE-9555 noting side marked TOP.

6. Replace carburettor bowl on carburettor barrel and secure by two screws, Y-110865, inserting filter Y-110576 under screw.

7. Replace air adjusting screw Y-110868 and spring in its socket, screw home gently and then turn back three quarters of a turn.

8. Replace air cowl Y-110860 and secure by screw Y-110861.

9. Replace gasket Y-9447 on manifold. Replace carburettor on engine and secure by two nuts and washers holding carburettor to manifold, using wrench Y-853.

10. Connect up carburettor controls and petrol pipes as in Section 1.D, Operations 7 and 10.

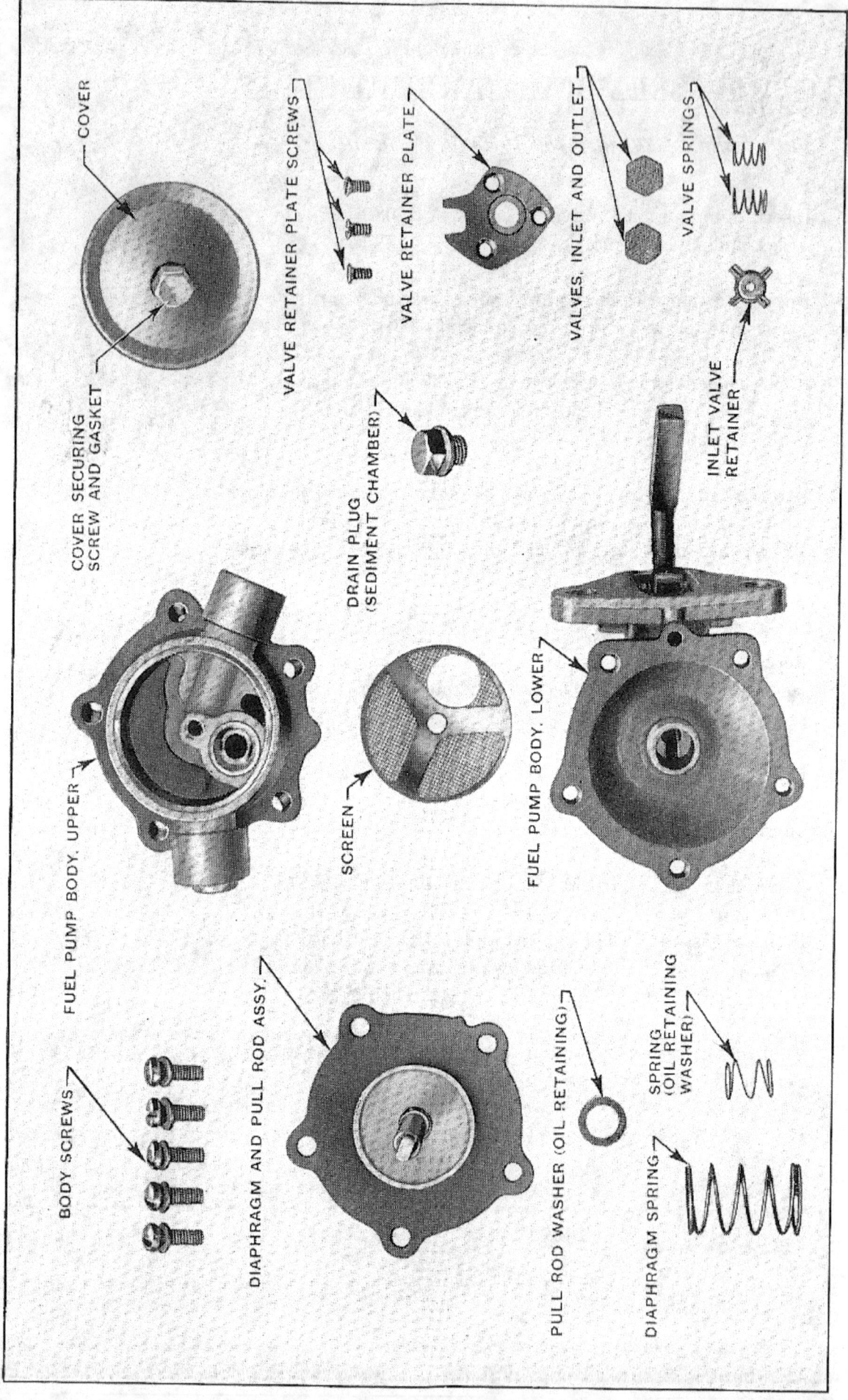

TO DISMANTLE FUEL PUMP

Carry out in sequence the following operations :—

Operation   Illustration

1. Remove flexible petrol pipe connection B-9288 at fuel pump, using wrench B-17015 and screw special cap on to flexible pipe to stop syphon effect from tank.

2. Remove nipple connection from pipe YE-9369-D to fuel pump with wrench B-17015.

3. Remove two bolts and spring washers holding fuel pump to engine using wrench Y-853 and lift off fuel pump and gasket Y-9374.

4. Remove screen cover YE-9355 by removing screw A-119020-S7 and washer YE-9357 with wrench Y-810. Preserve cork gasket YE-9364 carefully.

5. Remove screen YE-9365.

6. Remove five screws 31612-S2 and washers holding petrol pump body upper YE-9353 to petrol pump body lower YE-9375 and separate two halves of body.

7. Diaphragm and spring assembly can now be released from lower body YE-9375 by giving one quarter turn.

8. Remove three screws holding valve plate Y-110944 to body upper YE-9353. Lift off plate Y-110944 and gasket Y-110945. Remove two disc valves B-9358, two springs YE-9360 and inlet valve retainer Y-110943.

9. Remove sediment drain plug YE-9183 and washer, using wrench B-17015.

TO RE-ASSEMBLE FUEL PUMP

Carry out in sequence the following operations:—

Operation   *Illustration*

1. Wash all parts in petrol and dry thoroughly.

2. Replace sediment drain plug YE-9183 and washer, using wrench B-17015.

3. Replace one disc valve B-9358 on its seat on plate Y-110944. Replace other disc valve B-9358 on its seat in petrol pump body upper. Place two springs YE-9360 in their housings in petrol pump body upper. Place inlet valve retainer Y-110943 on top of spring resting on valve B-9358 on body upper. With springs as above replace plate Y-110944 and gasket Y-110945 on top of springs, so that three screw holes on plate register with holes in body upper. Replace three screws 25841-S and tighten down.

    NOTE.—To hold disc valve B-9358 in position on plate Y-110944 when replacing plate, very lightly smear with clean engine oil.

4. Replace oil retaining spring C-110324 and washer B-12213, and diaphragm spring YE-9396 on pull rod.

5. Place diaphragm and pull rod assembly in housing in petrol pump body lower and engage in slot in link Y-9381 by one quarter turn.

6. Fit two halves of petrol pump together with blind lug on body upper to rocker arm and secure by five screws 31612-S2 and washers.

7. Replace screen on top of fuel pump body. Replace screen cover YE-9355 and gasket YE-9364 and secure by screw A-119020-S7 and washer YE-9357, using wrench Y-810.

Operation *Illustration*

8. Insert two bolts 20346-S and washers into lugs on fuel pump. Place gasket Y-9374 on these screws and replace fuel pump on engine. Tighten down, using wrench Y-853.

   NOTE.—When replacing fuel pump on engine, care should be taken that rocker arm of fuel pump is not inserted under camshaft, but presses against its operating cam.

9. Remove special cap from flexible pipe connection B-9288. Fit flexible pipe connection in fuel pump intake socket and tighten down, using wrench B-17015.

10. Replace nipple connection from pipe YE-9369-D to fuel pump using wrench B-17015.

Page 30 "DE LUXE" AND "POPULAR" MODEL BULLETIN

GENERAL SPECIFICATION "DE LUXE" FORD

Fig. 28

Engine:

- Bore— 2.5 inches (63.5 mm.).
- Stroke— 3.64 inches (92.50 mm.).
- Capacity— 71.55 inches (1172 c.c.).
- R.A.C. Rating 10 h.p.
- Piston— Aluminium alloy—split skirt type.
- Piston Rings 3. Lower ring oil control—slotted type.
- Piston Pin Fully floating.
- Carburettor— Down-draught.
- Lubrication— Force feed to camshaft, main and big end bearings, splash to pistons, cylinders and piston pins. Valve stems lubricated by oil vapour from crankcase.
- Oil capacity— 5½ pints.
- Ignition— Fully automatic advance control.
- Sparking Plugs 14 mm.

- Cooling— Thermo syphon.
- Petrol Feed— Mechanical Pump driven from camshaft.
- Firing Order 1, 2, 4, 3.

Transmission:

- Gearbox— Similar to the 'Popular' Model Ford. Capacity 1¼ pints.
- Rear Axle— ¾ floating, 5.5 to 1 ratio. Spiral bevel drive. Capacity 1 pint.

Front Axle:

- Castor— 8 degrees.
- Camber— 2 degrees.
- Toe-in— $\frac{1}{16}$ inch to ⅛ inch.
- Front Spring Mounted in front of axle on forward extension of the front radius rods.

Specifications & Maintenance - Model C

"DE LUXE" AND "POPULAR" MODEL BULLETIN — Page 31

Wheels:
 Type— Hidden nut type, drop centre.
 Tyres— 4.50 × 17 angle type valve.
 Pressure— 35 lbs. per square inch front and rear.

Steering:
 Similar to 'Popular' Model Ford.
 Ratio— 10 to 1.
 Steering Wheel—16 inch diameter.

Brakes:
 Type— Improved internal expanding, 2-shoe type.
 Footbrake— Operating on four wheels.
 Handbrake— Operating on rear wheels.
 Diameter of drum— 10 inches.
 Width of drum 1¼ inches.

Chassis:
 Cross Members Three heavy-duty. Two of channel section, one of box section. Additional bracing by diagaonal gussets at front and rear.
 Side Members— Double drop reinforced at shock absorber mounting.
 Shock Absorbers—Hydraulic, similar to the 'Popular' Model Ford.

Electrical Equipment:
 System— 6 volt 61 ampere hour earth return.
 Lamps— 2 Headlamps bar reflector type, two wing lamps, combined stop and tail lamp. Instrument panel lamp controlled by separate switch.
 Direction Indicators—Sunk fitting type controlled from switch in gear lever knob containing tell-tale light. Switch operates through ignition switch to prevent unauthorised use.
 Battery— 6 volt.

Fuel Sytem:
 Tank Capacity—6½ gallons.
 Pump— Mechanically operated.
 Gauge— Electric. In circuit with ignition switch.

General Dimensions:
 Overall length (including bumpers)— 12 feet 1½ inches
 Overall Width .. .. 4 feet 9 inches
 Overall Height .. 5 feet 3 inches
 Ground Clearance (approximately) 8¼ inches
 Wheelbase .. .. .. 90 inches
 Track .. .. .. 45 inches
 Turning Circle (right and left) .. 33 feet

LUBRICATION AND MAINTENANCE

'DE LUXE' MODEL FORD

The importance of proper lubrication and periodic inspection and adjustments cannot be over-emphasised. The lubrication and maintenance work can be divided into two groups: first, points requiring attention every 1,000 miles; second, points requiring attention twice yearly or every 5,000 miles (whichever occurs first).

The lubrication chart, Fig. 29, gives information for complete lubrication. Proper lubrication has a vital effect on the life of any machine, consequently these instructions must be followed very carefully.

GROUP I.

AT 300 MILES, 1,000 MILES AND EACH 1,000 MILES THEREAFTER.

Engine:
Drain off the old oil when the new car has been driven 300 miles, and again when a total mileage of 1,000 miles has been reached and at each 1,000 miles thereafter. The oil will drain out more completely if warm, and should be replaced with approximately 5¼ pints of engine oil of the proper viscosity and quality.

The drain plug is on the outside of the oil sump immediately beneath the oil pump well in the sump.

Do not flush out the engine with paraffin.

Advise owners that oil level should be checked periodically between changes and recommend only oils of undoubted quality and of suitable grade.

Below are given specifications of suitable oils for winter and summer use:—

A neutral mineral oil, suitable for the lubrication of an internal combustion engine, of good uniform quality and free from deleterious substances. It must be free from acid, alkali moisture, tarry or suspended matter, thickeners or any other foreign matter, and should conform to the S.A.E.-30 viscosity range for winter use.

For summer use the oil should conform to the S.A.E.-40 viscosity range.

When recommending an oil to the owner, remember that cheap poor quality oils are never satisfactory, and eventually they will fail and re-act on YOU.

Chassis:
The chassis should be lubricated at 1,000 miles and at each 1,000 miles of operation thereafter. Suggest to the owner that the lubrication of the chassis and the changing of engine oil be performed at the same time.

Clutch Release Bearing:
The clutch release bearing is lubricated by means of a grease cup, located on the top of the clutch housing. The cup should be screwed in as far as it will go, then backed off and re-packed with a good grade of grease gun lubricant and replaced, screwing it in 2½ to 3 turns.

Generator:
The bearings are lubricated through a small hole at each end of the generator. Fill these holes with oil but take care not to over-lubricate, particularly at the rear, as excess oil is likely to work through on to the commutator and affect the operation of the generator.

Distributor:
Fill the oil cup at the side of the distributor with engine oil. A light film of vaseline should be applied to the cam.

Gearbox and Rear Axle:
Sufficient gear oil should be added to bring it level with the filler hole. A good quality oil should be used, conforming to the S.A.E.-160 viscosity for summer use, and the S.A.E.-110 viscosity range for winter use.

Universal Joint:
The universal joint housing should be filled with a special universal joint lubricant. A grease gun lubricator fitting is provided.

Grease Gun Fittings
Force grease gun lubricant to all parts equipped with the conical-shaped lubricator fittings (except universal joint)

Follow the chart carefully on all matters appertaining to lubrication with the conical-shaped grease gun fittings.

Be careful that no grease or oil is allowed to get on any rubber bushing.

Springs
The springs should be sprayed with a penetrating oil.

Fuel Pump
Drain sediment from fuel pump by means of drain plug.

Apply a Few Drops of Oil
Door hinges and locks, bonnet hinges and clips, spring tie bolt, accelerator cross shaft, and brake rod clevises.

Tyres
Air pressure in tyres should be checked and sufficient air added to bring the pressure to the recommended amount. Unequal tyre pressure results in uneven braking action and hard steering. Correct pressure is 35 lbs. per square inch.

Radiator
Water in the cooling system should be checked and replenished if required. (Radiator should be flushed at least twice yearly.) In winter anti-freezing solution should be checked for strength.

Battery
Inspect the battery and add sufficient distilled water to bring the electrolyte ⅜ to ½ inch above the tops of the plates. A rapid loss of water in the battery usually is an indication of an excessive charging rate, which should be corrected.

Lights
Inspect the various lamps and replace any bulbs necessary.

Axle Shaft and Wheel Nuts
Axle shaft and wheel nuts should be tightened after the first 300 miles of operation.

Cylinder Head Nuts
After the first 300 miles of operation the cylinder head nuts should be tightened. After this tightening they will require no further attention unless head is removed.

Carburettor
After the first 300 miles of operation clean and adjust carburettor.

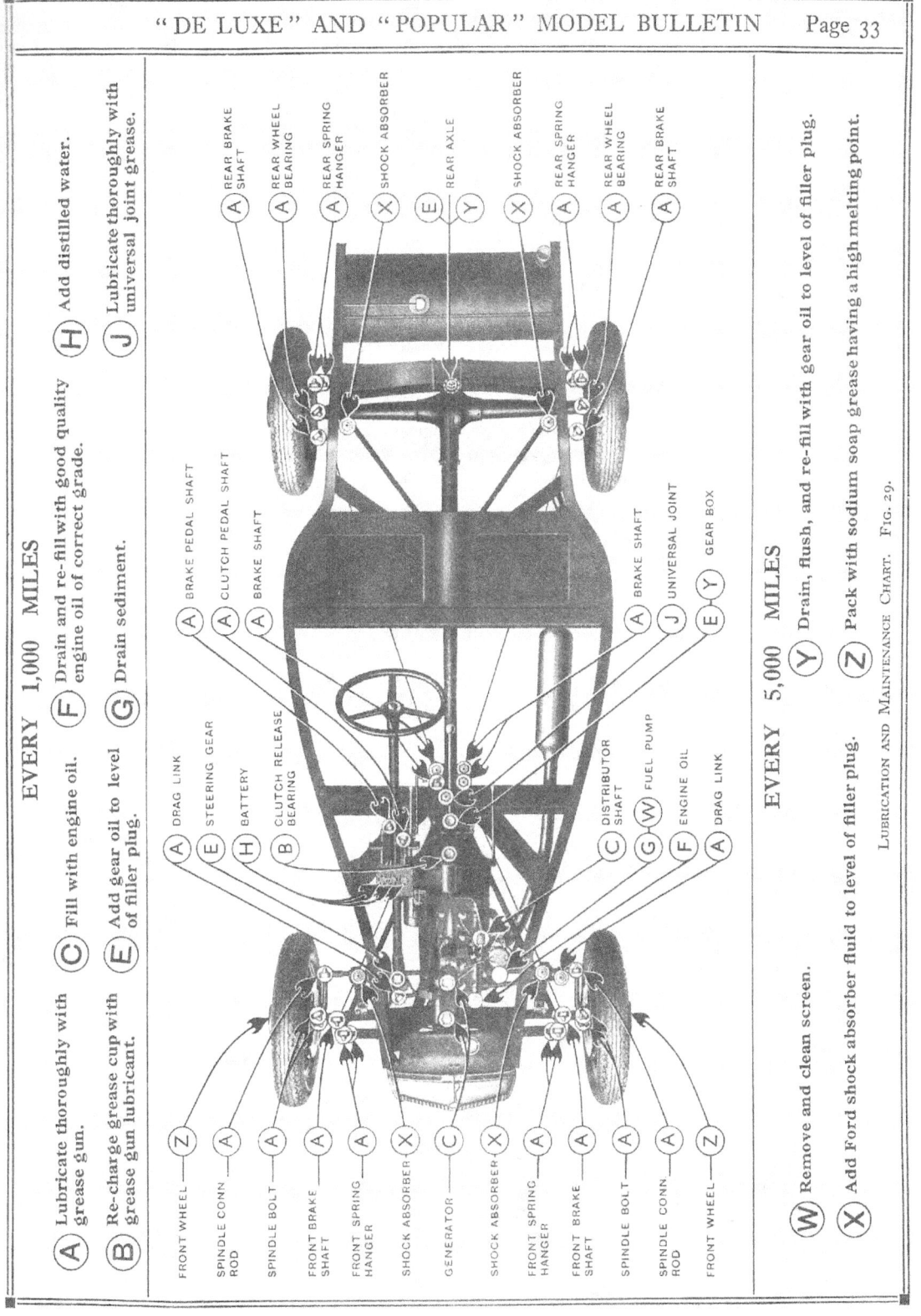

Lubrication and Maintenance Chart. Fig. 29.

GROUP II.

TWICE EACH YEAR, PREFERABLY IN THE AUTUMN AND SPRING, OR EVERY 5,000 MILES (WHICHEVER OCCURS FIRST).

In addition to all the lubrication and maintenance operations in Group I., the following operations are required.

Gearbox

The oil should be drained and the housing flushed with paraffin. Fresh gear oil should then be added until it reaches the level of the oil filler hole in the housing. Use the correct grade of oil. (See specification above.)

Capacity: 1¼ imperial pints.

The oil should be drained from the drain plug placed on the underside of the gearbox housing, and the housing should then be flushed out with paraffin.

To re-fill with fresh gear oil, remove the level plug, which will be found on the off-side of the gearbox. Then remove the gear lever by slacking off the locking ring immediately below the knurled cap which holds the gear lever to the selector cover. The cap may now be unscrewed and the gear lever removed. Pour the oil in until it starts to run out through the oil level hole. Replace the plug and gear lever and lock the knurled cap in position.

Always use the correct grade of oil (see specification above).

Never add oil to the gearbox without first removing the level plug, as an excess of oil will tend to leak forward into the clutch housing and cause clutch strip.

Rear Axle

The rear axle drain plug is situated at the bottom of the differential housing, the oil filler and level plug facing rearwards.

The oil should be drained from the drain plug and the housing flushed with paraffin.

Fresh gear oil of the correct grade should be added to the level of the filler plug.

It is important not to overfill the rear axle, as excessive oil will tend to leak into the brake drums and decrease the efficiency of the brakes.

Capacity: 1 imperial pint.

Front Wheels

Twice yearly or every 5,000 miles (whichever occurs first), or at any time when the car has been operated with a front wheel hub cap missing, the front hubs should be removed and the bearings and the inside of the hub washed clean with paraffin and re-packed with a short fibre sodium soap grease having a melting point of not less than 350° F.

Shock Absorbers

The level of the fluid in shock absorbers should be checked and sufficient fluid added until it reaches the level of the filler plug. Only genuine Ford shock absorber fluid should be used.

Ignition

Inspect the gaps between the contact breaker points as well as the sparking plug gaps and adjust as required. This attention should be given earlier if misfiring is evident.

Battery

Inspect battery connections and clean if corroded.

Starting Motor

The bearings in the starting motor are lubricated upon installation, and require no further attention.

Clean the commutator by holding a strip of very fine glass paper against it with a small piece of wood while the starter is operated. Blow out any carbon dust that may have accumulated, examine brushes for excessive wear and see that all cable connections are clean and tight.

Spring Clip Nuts

Inspect these nuts and tighten if necessary.

Clutch

Check the amount of free travel of the clutch pedal and adjust if required.

Brakes

Check the movement of the brake pedal, re-adjusting the brakes if the pedal travels to within two inches of the floor board when the brakes are applied.

Fuel Pump

Clean the fuel pump screen.

Generator

Adjust charging rate to conform with owner's requirements.

Clean the commutator by holding a strip of very fine glass paper against it with a small piece of wood while the engine is idling. Blow out any carbon dust that may have accumulated and examine brushes for excessive wear.

Replace if they have worn to such an extent that there is any possibility of the brush lead fastenings fouling the commutator, or if excessive sparking occurs. See that all connections are clean and tight.

Body

Suggest to owners that a periodical application of body polish will enhance and preserve the lustre and beauty of the body and mudguards.

www.ingramcontent.com/pod-product-compliance
Lightning Source LLC
Chambersburg PA
CBHW060415220526
45465CB00008B/2896